Gerhard Oberniedermaier
Tamara Sell-Jander

Sales and Distribution with SAP®

vieweg IT

The Efficiency of Theorem Proving Strategies
by David A. Plaisted and Yunshan Zhu

Applied Pattern Recognition
by Dietrich W. R. Paulus and Joachim Hornegger

SAP® R/3® Interfacing using BAPIs
by Gerd Moser

Scalable Search in Computer Chess
by Ernst A. Heinz

The SAP® R/3® Guide to EDI and Interfaces
by Axel Angeli, Ulrich Streit and Robi Gonfalonieri

**Optimising Business Performance
with Standard Software Systems**
by Heinz-Dieter Knöll, Lukas W. H. Kühl,
Roland W. A. Kühl and Robert Moreton

ASP – Application Service Providing
by SCN Education B.V.

Customer Relationship Management
by SCN Education B.V.

Data Warehousing
by SCN Education B.V.

Electronic Banking
by SCN Education B.V.

Mobile Networking with WAP
by SCN Education B.V.

Efficient eReporting with SAP EC®
by Andreas H. Schuler and Andreas Pfeifer

Interactive Broadband Media
by Nikolas Mohr and Gerhard P. Thomas

Sales and Distribution with SAP®
by Gerhard Oberniedermaier and Tamara Sell-Jander

www.vieweg .de

Gerhard Oberniedermaier
Tamara Sell-Jander

Sales and Distribution with SAP®

Making SAP SD®
Work for Your Business

vieweg

Die Deutsche Bibliothek - CIP-Cataloguing-in-Publication-Data
A catalogue record for this publication is available from
Die Deutschen Bibliothek.

1st edition June 2002

Vieweg is a company in the specialist publishing group BertelsmannSpringer.
www.vieweg.de

Cover design: Ulrike Weigel, www.CorporateDesignGroup.de
Printing and binding: Lengericher Handeldruckerei, Lengerich
Printed on acid-free paper.
Printed in Germany

ISBN 3-528-05770-X

Table of content

10

Data archiving in the module SD 211

15

16

Preface

Today's business environment is shaped by several factors. Especially important are permanently growing competitive pressures and continuously increasing customer demands. Within the changing environment it is essential to know and to optimize your company specific business processes to meet those demands most efficiently.

Efficient process design gained a high importance within the last years. Because of the variety of variants that can be implemented with the SAP R/3-system, a profound knowledge of the business background is necessary to use its full potential.

The intention of this book is to provide the interested reader with applicable solution sets to implement the SAP R/3 SD module.

A more detailed description of the motivation and content can be found in the following introductory chapter.

Clearly we are always grateful to all our readers for their comments and suggestions regarding the content and structure of this book. We expressively encourage readers to send us constructive criticism and suggested improvements to this current, first edition of the book. We will do our utmost to respond to their suggestions.

As I am the primary author please use the following e-mail address for further contact:

mailto:Gerhard.Oberniedermaier@t-online.de

Readers can use the same address to direct their inquiries about the implementation of R/3 projects to me.

Here I would like to thank all those who have contributed to the successful completion of this book.

Mühldorf a. Inn, Germany, April 2002

Gerhard Oberniedermaier

Introduction

Logistics signifies the structure of material, information and pro-duction flows from the deliverer over the production to the consumer. In the second half of the 80's the term Logistics developed itself more and more to a slogan.

Even in the future, most of all on grounds of the advancing globalization of enterprises and the world economy, logistics will increasingly gain significance for the saving of the enterprise's success. On the one hand logistics enables the enterprises to gain considerable advantages in competition, in ever more aggressive markets, by the use of innovative and progressive structure of the material and information flow. On the other hand the powerful pressure of cost permanently forces into the increase of efficiency.

For various studies substantiate, that the part of logistics cost shared in the total expenses ranges in the scope of 10-25%, which fundamentally affects the situation of results of an enterprise. Therefore the maximal necessity is to concern oneself intensively with the logistics way of looking at a problem and with that to search for ways to optimize the material and information flow with the logistics chain.

At today's markets the pressure of competition of enterprises is nearly permanently growing. The one who wants to be able to meet demand of ever more global markets, must not only gain control of one's cost but also possess the ability to reduce these lastingly. Therefore in almost every enterprise today one talks of the so-called Supply Chain Management, which aims the goal to make processes of logistics as such more flexible, efficient and inexpensive.

Should a solution like this imply a real advantage, the various experience show, that not only the use of certain software products is necessary. The changes must be placed much deeper in position and stretch from the alteration of the processes up to a completely new thinking on part of the employees on the enterprise.

To reduce costs of material flow, the different units, from sales over production to purchase, must be connected more tightly. The acceleration and optimizing of processes is in this way, in

3

common sense only possible through Supply Chain Management. Besides the reduction of costs, more efficient structures of logistics offer even more improvements in the capacity of supply and in the reliability and next the higher customer satisfaction. Further on the length of the run can be decreased deliberately.

In order to consist on today's markets or keep permanently competitive, the enterprises must change the often traditionally grown and therefore often antiquated structures. The slogan *Business Reengineering* is on everyone's lips.

Every enterprise tries hard to process it's own organization, the production and naturally also the products according to these requirements of the market. The successful enterprise of the future will therefore differ from today's enterprises in many ways. Naturally these differences will extensively effectuate the whole field of information technology. With Business Reengineering and information-technological renewals standardized software plays an important role on the way to the era of information technology. An integrated "Standardized application" software includes various impulses for a quicker and, compared with the quality of information, improved processing of information within operational business processes. As examples may be mentioned here:

- Saving of time by automatic creation of order proposal

- Automatic posting of payments receipt versus open items

- Automatic posting of goods receipt versus an open purchase order and automatic posting of the printing of purchase order fixed in the G/L account

- The combination of amount –and value movements: Therefore it is not necessary to transport a specific receipt to the accounting, but the balance sheet account is automatically increased by the goods receipt

The modern enterprises recognize continually, that on today's competition they can only exist, by speeding up their business processes. An elementary requirement for this however is, that traditionally separated processes –from sales and marketing up to production and supply- get completely integrated. This for example can be realized during and with the implementation of the operational standardized software SAP R/3. SAP R/3, with its option of the enterprise specific extension through the existing surrounding of development, phases the reorganization of the enterprises, even of the ones of the middle-class, on 'future safe' ground on the way into the information society.

For the moment it is not relevant, in what way the enterprise approaches the subject of business reengineering as a project which lays before the introduction of the standard software or – which may be rather preferred under the aspects of cost and time- as a continual process of improvement after the introduction. [Jakob/Uhink98/25]

Even the already mentioned slogan 'Supply Chain Management' must not be forgotten here. More and more branches recognize the advantages, which come along with Supply Chain Management. In the chip production and in the metal industry it nearly is already obvious. The automobile industry immediately faces a supply chain management revolution. Further branches of industries will follow.

For more and more enterprises organize their 'frontier crossing' business by building up a universal chain of logistics. This integrated supply chain forms a continuous process from the raw material production, the production over the distribution systems, up to the customer. This implies an increasing speed of the globalization of the world economy, which forces all enterprises to quicker action, for not missing the connection by losing grip on the new markets, new technologies and new changes. Information systems, for an efficient business process within the scope of logistics systems, nowadays represent a further factor of competition with enormous importance.

Both the planning and the supply and distribution must be adjusted to the processes of manufacture. The smoothing procedure of the logistics chain generally depends on technical possibilities for satisfaction of demands. In order to gain the most possible advantage with the Supply Chain Management techniques, the production enterprises must work on optimizing their processes of manufacturing. Besides it's not about accelerating the processes, but making them more efficient. The production procedure differs from branches to branches. Unlike the accountancy there are no common rules for the production (every branch is provided with production processes of it's own, which have been developed on base of optimal processes, cost structures and demands of competition. In order to obtain higher profit margins the producer of chemical goods for example, have to optimize their manufacturing process. For enterprise' s low influence on the prices of the raw materials used, profit can be obtained merely with proportionate production and optimized production control. An enterprise of the electronic branch must otherwise have the skill to restructure its manufacturing processes rapidly, for adjusting them to the frequent construction changes and technical improvement. In this field there are rules for fixed retail prices for the components, and the first enterprise, which puts a technical innovation on the market, will win the fight for the shares of the market and obtains the highest profit margins. Though there are not only differences between the branches, but also in under-segments of a branch. The processes and demands of the automobile manufactures for example differ elementarily from those of the system suppliers or component supplier. And the processes of production of system vendors producing drive shafts, differ elementarily from those producing seats or electronic parts. Therefore the requirements do not only vary considerably by branch, but also by segment and the production type. Although some basic components are necessary for all producers, there are differences depending on the branch, which have to be considered in the manufacturing processes. Therefore the standard software, coming into questionnaire within the limits of Reengineering and Supply Chain Management must be outlined in such way that it supports as well the general as the branch specific functions. For that purpose the success of an application is determined by knowledge and a broad functionality. Information and product flows within the whole market hide behind the concept Supply Chain Management. Various studies of Newbury Port, an institute of market research belonging to the Industry Direction Inc. in Massachusetts, have shown, that the bottle neck in the information flow, both within the single companies, as well as between companies in a logistic chain have far reaching consequences on the efficiency of the enterprises. Information pushes the transmission and therefore the flow of products with the help of the logistic chain. This requires the coordination of various different activities at different places. In the semiconductor industry it is estimated, that at least 1 billion Dollars could be saved per year by eliminating informational bottlenecks of information.

At present it becomes very often clear that the technology and supporting systems of an enterprise are not easily transformable to another enterprise. Specific surroundings and various factors decide, what functions are necessary for an enterprise. Industry Directions Inc. has concentrated on the evaluation of process strategies and software products, required for special branches and procedures. The examinations result in the fact that the suppliers of software must either specialize on a few branches or complete general available basic functions. With 'branch specific' functions even the permanently growing importance of Internet contributes to the intensification of the market. The Internet becomes more important, because of internationalization of markets, the globalization, an increasing mobility of capital and the fact that information has grown to one of the biggest economic factors, and therefore processes high potential of changes in view of operational processes. The logistics chain can be extended with help of the Internet by integration of suppliers and customer. The Internet functions as medium for the flow of information, which controls itself, and serves for the coordination of the attitudes of the various participants. In former years moreover, the change from supplier to customer market was influenced by the change of the logistics chain, so-called Supply Chain, to a great deal.

Since the change from sales to customer market [Nieschlag94], the sales becomes a permanent growing challenge for the suppliers, because it is not enough, on such markets, to merely distribute goods, but sale procedures in an active professional way become necessary. The sales departments nowadays are under great cost pressure. Besides the aim of logistic to working more inexpensively, it should be worked more rapidly and specifically, that is more customer-orientated. But this can only be realized by a forced use of new information technologies. On grounds of the new sales potential via Internet, which is similar to the mail order business, many tasks can be taken over by the customer itself, with help of online-order formulas.

The markets nowadays are increasingly marked by a more intensive competition. The customers' demands on a product therefore get more unstable and can't be planed and prognosticated. In the first place today's sales prospects are specified of how exactly the customers´ conceptions are covered up within the framework of their demands. Today's enterprises of most branches face this challenge by permanently rising focus on the respective market and the single customer. Thus, that industrial structure, in which product programs of the big enterprises could have a great influence on stabile market, are becoming more and more obsolete.

Nowadays the push-principle is more and more replaced by the pull-principle. Within the pull-principle market demands control the supply of the customer via the logistics chain. That means a permanently fluctuating, varying of demand for every enterprise, for the customer demand quality cannot be planned 100%. The result is, that either the production as well as the distribution of every enterprise must be able to flexibly face changing demands and therefore at least to minimize the risk of full stocks and consequently high costs.

The discussion about the goal and use of the pull-principle is not new .The pull-principle has been essentially influenced by the conception of the 'lean production', the 'Kanban Control' and the just-in-time management basis.

The Supply Chain itself describes the whole flow of goods from the first supplier to the customer. Following partners are involved in this process:

* Supplier

* Producer

* Distributor

* Trade

* Consumer

As one recognizes, a lot of different business partners are involved in the process of the logistics chain, as for example, partners of own enterprises, to which external market relations do exist. Thus the elementary challenge within the scope of the Supply Chain Management lies in the comprehensive harmonization of the information and goods flow along the logistics chain. This has particular validity for global operating enterprises, which have extended product compound systems into an internal Supply Chain. New business comprehensive information systems now form a strategic part of the supply chain management.

These operational information systems have up to now confirmed within the limits of an enterprise to the processes of the Supply Chain, which have to be scoped with by every participant of the logistics chain. In the near future one will not deal any longer with isolated systems for closed business units, but with comprehensive systems, which rule the transfer of goods and services as well as the payment flow through the restricted information interchange.

SAP has gone the last way and has first of all developed a solid score application with all-important basic functions and later on extended them with vertical specific functions, components and initiatives. A general production flow with required manufacturing processes and software functions are the key to success. A powerful software and the integration of ERP manufacturing functions and specific controlling application guarantee the smoothing and interruption free procedure of the logistics chain. The capability to rapidly integrate new technologies, to transform these to useful solutions and make these prematurely available to the customer's advantage, belongs to the key to a successful standard user application. The success story of the system SAP R/3 is essentially based upon these attributes.

The letters SAP stand as abbreviation for System, Application and Products in the data processing. Goal of an enterprise was and is, to develop and sell a single operational standard application for enterprises, which covers all operational units and offers the customer a standardized structure and a graphical user interface.

This application system then should not merely function in one single branch, but further on in the whole range of all industry branches, as well as in the service sector and public services. The application neutrality should even go beyond the borders of a country, so that the system would be internationally employed. This concept was exceptionally successful. Even 1980, this 8 Years after the company was founded 50 of the 100 most successful German industry enterprises belonged to its customers. In the meantime there are 95.

Nowadays the R/3 System has become a de-facto standard system for the big, multi nationally operating enterprises as well as for smaller, nationally engaged companies.

With the implementation of user modules offered in the SAP R/3 system, logistics procedures can be planed on grounds of existing data- and function integration. The integration of the single user modules with R/3 prevents unnecessary and time intensive multi entries by dealing with logistics transactions. Further on the ad value side is registered within the quantitatively orientated processing steps, with which the requirements of the accounting can be taken into consideration. In particular sales, materials management, production planning, quality management, project system, service management and maintenance belong to the user modules of logistics.

The book at hand is structured as follows:

In chapter 1 of this book is briefly described:

- Definition of the term logistics
- Targets and tasks of logistics
- Division of logistics in the sections sales logistics, supply- and production logistics
- Description of the tasks of these sections
- Explanation of an information-technological realization of these tasks

This should offer a complete view to these readers, who have up to now no specific knowledge of these themes.

- In chapter 2 are briefly described:
- The term standard application
- The features of standard application
- SAP R/2 and R/3
- Procedure models for the implementation
- Recommendations for a successful implementation

In chapter 3 logistics modules of SAP R/3 and their functionality are briefly and precisely represented. Following modules are herewith described:

- Material management
- Production planning
- Quality assurance/management
- Maintenance
- Service management
- Project system
- Sales

Chapter 4 describes the efficiency characteristics of the R/3 logistic modules SD in regard of the operational requirements on sales logistics. Subsequently, the most frequent operational scenario for the realization of an efficient Business Reengineering of the sales logistics scenarios with the module SD are described in detail in chapter 5. There are for example.

- Order processing
- Third-party order
- Management of consignment stocks
- Etc.

In chapter 6 the master data as basis for all processes is represented. They are the operational basic data of an enterprise. Their quality is the fundamental basis for the successful procedure of transaction.

Chapter 7 describes the functionalities of pricing in the module SD, which are:

- Price conditions
- Condition methods
- Condition types
- Condition tables
- Access sequences
- Pricing schemes
- Determination of pricing schemes

Chapter 8 shows the use of the sales information systems. This information system makes both the management and the sales employees compact and extremely predicative information available. This information on different summarizing levels makes the recognition of changes on the market possible. It forms the fundament in order to make strategic and operative decisions.

An efficient sales controlling is supported by information functions and analysis referring to the most important transactions.

With the availability check in the module SD there can follow a passing on of requirements. That means, the material requirements planning department/controller is informed about the amount required by the sales in order to be able to deliver the received orders. With the integration of the R/3-System the information change necessary between the applications sales (SD), material management (MM) and production planning follows automatically. The requirements will then be reported in form of single or collective requirements. An introduction in the thematic of the requirements planning is made in chapter 9.

Chapter 10 should help the reader with the planning and implementation of the necessary data archiving in the SD module. And it points out possible restrictions and dependencies.

Fundamental target of chapter 11 is, to bring transparency into the controversially discussed theme online-sales/electronic commerce (EC). Before this background the EC is watched operationally, in order to recognize chances and risks. Herewith it should most of all be referred to Business to Business (B2B) and Business to Consumer (B2C) by Internet, but also the basis of the implementation of sales-EC solutions in connection with SAP R/3.

The success of a standard application implementation depends on a methodical procedure. Understandably the enterprises, the one's of the middleclass, particularly want to keep the implementation time for new systems and new processes low. Nevertheless the quality and productivity of the new system should be on highest standard. This ambition is challenged by SAP prematurely by making, among others, qualified methods and tools available, which should help the customers to essentially reduce the implementation time. In order to meet the demands of SAP-customers on shorter implementation duration the SAP AG developed the implementation method ASAP-Accelerated SAP.

In chapter 12 ASAP, on basis of the continuous business engineering (CBR) developed implementation method, is represented. A questionnaire catalogue should serve for making an implementation of the module SD in accordance to the ASAP-methodology easier and arranging certain solutions.

As already described, in the former years the concept of the Supply Chain Management (SCM) has become an important concept of the management, which receives great attention. For that reason both, the task and the strategically and operative targets of SCM, should be described in chapter 13. Afterwards the realization of the SCM in sales for the departments' order-entry, creditworthiness checking, delivery check and e-commerce is described. Additionally, in this chapter the transformation of these SCM-strategies in the SAP surrounding is looked at. In scope of its Supply Chain Management initiative, which belongs to the new dimension of SAP, SAP developed various tools. APO-Advanced Planning Organizer, B to B-Procurement and the Logistics Extension System. These tools and their significance for the sales are also explained.

The target of chapter 14 is, to give an overview over the CATT-Computer Aided Test Tool-within the implementation of the R/3 component SD –sales and distribution- in accordance with implementation methods (traditional implementation, ASAP-Implementation methodology, DSDM based procedure model) known. With that, a practice orientated guide should be offered to both SAP consultants and users, which should precisely explain the application possibilities of CATT during the implementation and even the optimization and further development of the SD module.

The practice orientated case study in chapter 15 shows how invoicing -with the appropriate commission calculation- can be examined and automatically be transferred to a commission demand.

In chapter 16 we define in scope of a case study the controlling of the (revenue) accounts by the system to the handover of invoicing elements/data from the module SD to the module FI. The account determination takes place with the help of condition techniques. The corresponding procedure of customizing is described.

Appendix A describes the traditional procedure model for the implementation of R/3.

Appendix B shows all working packages of the ASAP-implementation method.

It may be emphasized that the book at hand treats the thematic SD in such way as that the R/3 release versions 4.5 and 4.6 are covered.

1 Logistics

1.1 Definition

As already mentioned, logistics is understood as the application of the material-, information- and production flow from the supplier over the production to the customer. In the second half of the 80's this term developed more and more a slogan.

Logistics is a term derived from the Greece 'lego'. It means thinking, calculation and considering. About the year 1780 the term logistics gained importance within the military. The logistics there covers the planning, supply and the deployment of required military means and services for the support of the forces [Bartels 80].

In the beginning 60's the, in the field of military aspects, logistical experiences made in the USA are transferred to economics. The economic development of this century, which is marked by a tremendously growth of enterprises and an expansion on different markets, caused the compulsion to coordinated and controlled movement of all material and goods flow. With that, logistical considerations entered these companies whose tasks have in the meantime expanded over the whole basic function chain from the purchase over the production to sales and distribution.

Both fields have in common the transport of goods. Though the term logistics in the field of military includes troops and goods, whereas in economics exclusively goods. The second difference is that the military logistical decisions are orientated on political and military targets, whereas management logistical decisions lie on economical targets [Keller99].

The economy intensively examined this appendix. To the department of logistics, processes of an enterprise, which are concerned with materials, are to be counted. The examination of logistics is discussed within the decision and system orientated appendences of the economies. There, enterprises and their environment looked at as complex systems and decisions for the application control and regulation of economics systems analyzed [Pfohl 80].

The economic logistics has the target to achieve the economic application, regulation and control of the material and information flow of an enterprise. The application includes the relationship from the supplier over the enterprise to the consumer. In the broad sense the task of logistics is to secure the availability of goods at a specific place to a specific time with the right amount. In the narrow sense the logistics tasks cover the transportation, storage and arrangement of materials [Bloech 84].

Starting point of the practical usage of logistics in enterprises was the supply of the production with material, in the seventies. Today the term logistics comprises the regulation of activity performance processes and includes the coordination of the material-, information-and production flow beyond its duties. In general, logistics comprises the duties of performance:

- Delivery relation ship to the customer

- Procurement relation ship to the deliverer

- Production procedure with the planning of the transport routes and storage of semi-finished-and finished products.

Logistics has, in the second half of the eighties, become a buzzword and a sparkling term. A corresponding term background is only hardly determined, with different authors and associations.

Logistics is considered a market orientated, integrated planning, performance, procedure and control of the entire material-and corresponding information flow between companies and their deliverers, within a company as well as between an enterprise and its customers [Schulte95/1].

In regard of the exact knowledge of these terms it is necessary to compare the mentioned terms and differentiate, on the one hand the functions allocated to the terms and on the other hand the objects to be looked at through them.

1.1.1   Objects of logistics

As objects of logistics shall be regarded all materials and goods, which are manufacturing goods, process materials and operating supplies, components and spare parts, commercial goods, semi finished and finished products as well as remaining goods.

1.1.2 Targets of logistics

Target of every logistical activity is the optimization of logistical profit with its components logistical performance/- costs.

Logistical performance

Elements of the logistical performance are:

- Delivery time
- Delivery reliance
- Delivery flexibility
- Delivery quality
- Information capability

Logistical costs

The second component of the logistical profit is formed by the logistical costs, which can roughly be divided in five cost centers [Roell85]:

- Control and system costs
- Inventory costs
- Stock costs
- Transport costs
- Handling costs

Empirical surveys discovered that the logistical costs have a part generally more than 10% on the total costs. This emphasizes the great importance of logistics for the result situation of an enterprise.

Optimization of the logistical profits

Roell fundamentally offers two ways for the optimization of the logistical profits:

- Realization of an optimal logistics performance grade
- Pursuance of a demanded logistics performance grade with minimization of the relevant logistic costs

Profitability factors of logistics

On behalf of the National Association of Accounts and the Council of Logistics Management a survey has been carried out, in which ten profit factors of enterprises with very good logistics could be identified [Busher/Tyndall87].

1. All aspects of logistical activities should directly be connected with the strategic business plan. The most important principle, which should be applied for the complete exhaustion of the profit increase potential with help of logistics, is the immediate connection of logistical activities with the business strategy. Logistics is hereby applied to support the achievement of competitive advantages.

2. All logistical functions should be fully organized. This principle means that all logistical functions should be unified in one organizational unit. Herewith an essential reduction of the decision ways is to be achieved and therefore the information lead times.

3. Successful logistics departments draw all benefits from the information- and communication technology. Successful logistics systems apply information as a strategic resource. The electronic data exchange for example can be part of a differentiation strategy and build up entrance barriers for the competitors.

4. Logistics adjusted to the personal politics, is a prerequisite for excellent logistics performance. It appears, that experienced, well-trained logistics managers are essential for the successful realization of logistics strategies.

5. Enterprises should cultivate a close partnership with other participants within the logistic chain. These relationships are not any longer regarded as short cost reduction programs, but strategic alliances are built up with deliverers, customers and forwarders, which partially even imply the incorporation into the planning of new projects. This in particular applies the implementation of the just-in-time principle.

6. Enterprises should call upon predicative indicators by the valuation of the logistics efficiency. It appears, that the functions transport, stocking, and shipment service can be controlled most efficiently, if they are organized as cost-or profit centers.

7. Enterprises, which meet the optimal service grade, improve their profitability.

8. Attention to detail can imply great cost savings. Here it also depends on the connection of the details with the business strategy. Thus a logistical strategy, e.g. which of-

fers the wrong precautions for the customer satisfaction, can lead to sales declines, which are ostensibly interpreted as operative problems.

9. Successful logistical systems consolidate transport volume, stocks etc. with the intention, to gain operative and financial depression effects. These aggregations often lead to evident service improvements and cost savings. Prerequisites for the achievement of these effects are however, a well-founded analysis of the available alternatives and trade-off relationships. The collection of consignments is, in this connection, the most often applied method. Various, on markets available, software systems for the optimization of the route plan support this starting point. A further starting for the consolidation of amounts lies in the reduction of the forwarders amount. Even the observed trend towards the centralization of the stocks, turns out to be very profitable on grounds of the stock reductions to be gained, despite of high capital expenditures.

10. Enterprises must measure their logistical performances and react in a dynamic, serial process on the results.

1.2 Functions of logistics

Traditionally the function sectors of logistics are divided in purchase-, production-, and operational logistics and sales logistics and distribution logistics [Pfohl96].

1.2.1 Sales logistics

The sales logistics is the link between the production and the sales part of the enterprise. Task of the sales logistics is, to perform, control and regulate the goods flow of an enterprise on or to the sales market. Target is, to deliver the demanded goods in time to the customers with the minimization of delivery and transport costs. Though it's all about the optimal operation of the chosen sales variations [Traumann76]. The sales logistics implies the activities from the demand and often over the order dealing and control, the shipping and the invoicing to the financial transactions, which check the proper entry of the payment. Depending on the business type, the sales logistics can appear as follows:

• It can be temporally neutralized from the procurement and production logistics e.g. anonymous stock manufactures

- It can directly be coupled with the production logistics, as e.g. a deliverer fabricating order related products out of stored raw materials

- It can directly be combined with the procurement and production logistics, as e.g. an automobile manufacturer, configuring his offered automobiles customer individually and whose procurement materials are delivered in accordance with the just-in-time procedure.

More and more enterprises apply the goods distribution besides other sales political instruments as a competition instrument, in order to achieve advantages through an improved delivery service with regard to the competitors. With that the necessity arises of taking into account the customer requirements, which more often lie in the overtaking of additional performances as stock keeping [Poth73].

The customers increasingly try to reduce their own stocks, by ordering the demand related synchronously smaller amounts in shorter intervals. This forces the delivering enterprises, to develop delivery strategies, which guarantee a high delivery preparedness but without leading to a cost explosion [Schulte95].

The most important problems of the sales logistics, which have to be faced, are:

- Order handling

- Order picking and packing

- Goods issue

- Shipment

Order handling

For the control of the whole goods flow within the goods distribution and the coordination of all single activities, the use of an order handling system is absolutely necessary. This makes possible the interaction of men, institutions and processes with certain structures, by providing qualified information. The order handling of a sales system is therefore superior to the operative processes of the product flow. With that the building up of different auxiliary systems, which perform certain tasks, as the entry of master data, the stocking, goods accountancy and so on, serves to acceleration and rationalization of the information flow procedure. The rating of such system is very high, since only through the availability of premature and comprehensive information quick and flexible sales can be realized on long-term. The application of information technologies in case of the order handling has been brought about in most enterprises. The partial-and full automation connected with that takes place on several levels:

- On lowest operational level, on which process-based systems are applied for data entries and data output

- On the second level, on which logistical partial systems are developed with the connection of single elements

- On the third level, on which the single subsystems are connected with each other in a network and are applied for system control, -disposition and the permanent planning

- On the highest level, which, as logistics information and control center, occupies with long-term capacity and application planning

It is, to plan the building-up of these systems in such a way, that expansion possibilities do exist, in order to guarantee a sufficient flexibility for future technological developments or acquisitions of the market. The application of modern communicative techniques leads to high investment for an enterprise and require in many cases structural organization changes. Therefore, with such a decision, an accurate survey of the profitability and requirements of the market has to take place. On ground of the high demand on faster provision of information, a data transfer without documents and a decoupling of the data flow from the goods flow is strived for. Further more a total integration of the system is to be gained. The necessary investments are opposed to far reaching cost reduction potential: They lie, among others, in an improved and more rationalized accountancy and cost accounting system, the optimization of route and stocking plans and a more precise adjustment of the cooperative partners' time plan as well as the undertaking of a lot of routine work through the PC. Further more, with electronic data interchange between shipping agent and consumer a higher dependency of the customer can be achieved, since a change to another vendor can imply a switch of the information system, which is often connected with high cost for new systems and the retraining of the employees. Any recourse possibilities on existent databases of the enterprise or its cooperation partners facilitate and accelerate the work process. Since the necessary hard- and software in most cases is implemented step by step, a modular building up of single subsystems, is very practical, in order to secure function capacity independently from each other.

Picking

Picking includes the arrangement of certain subsets out of a total set provided on grounds of requirement information. Hereby a transmission of a stock specific into a demand specific condition takes place. In general a store function is placed ahead of the picking and a consumption function (e.g. production, assembly, dispatch afterwards.

The single steps necessary for the picking often require a high coordination –and control expenditure. An enormous rationalization potential lies therefore in general in the procedure structure of the picking. It is made accessible by a material flow based relation of the warehouse-and picking functions, the optimization of the techniques used as well as the integration of the material and information flow.

Following basic functions are to be made in the scope of the picking (Schwarting86):

- Preparation of demand information (picking orders)

- Preparation of material groups

- Controlled take-out of subsets off the prepared total set

- Methodical progression for the take-out and delivery

- Delivery of subsets to instances and receipts of execution.

Goods issue

In order to secure short material and information lead times as well as low total costs in the field goods issue a variety of features are to be considered. The turnover and transport establishment necessary is most of all influenced by condition, shape and weight of the material leaving. The temporal distribution of consignment determines the personnel and transport capacities necessary. Size and shape of buildings, warehouse spaces, working spaces and offices depend on the type of the materials. Technical establishments are important for the interim storage and preparation, the material movements, the control transactions as well as the packing. With the help of organizational rules, tasks ranges and limitations are to be determined, the procedures to be controlled and the information and document flow to be specified.

Shipment

The shipment of materials and goods serves the overcoming of special distances. With the external transport in the scope of the sales logistics, the transport from enterprise to customer takes place.

The external shipment is one of those parts of the logistics, which are the most influenced by external conditions. Many of today's branches go through a change in structure, which is marked by following developments:

- Intensification of the world wide division of labor and globalization

- Transfer of work intensive production procedures to foreign countries

- Reduction of manufacturing "depths", that is, transfer of parts of the production to component suppliers

- Production in small lot sizes with the target of a plants' stock reduction (no capital lock up in stocks).

This leads to higher requirements on the transport because of

- Reduced order amounts

- Most frequent delivery and

- The demand on ever more qualitative and technically more complex products, which therefore show a higher sensibility towards burden.

1.2.2   Production logistics

The production logistics has the task to provide the optimal preparation of semi products into the single steps of the production process. The stress lies here in the physical structure of the material flow. Besides the essential production and assembly of component parts and assembly parts to a sellable product, following functions do also belong to, beside the already mentioned transport and storage steps:

- Creation of a material flow related factory structure (factory planning)

- Production control, that is the logistical procedure between the machines and the logistical control of the single machines

- Production planning with consideration of capacities, amounts and resources.

The higher the level of production automation developed by an enterprise, the more intensive is the operational logistics connected with the production.

While the enterprise planning is a medium to long-term effective decisions, which has to be connected with the strategic business planning, a medium to short-term time horizon underlines the production planning and control. As far as the choice and establishment of the implementation in the enterprise system of the production planning and control is concerned, this is even a structural decision with long-term planning horizon.

The planning and disposition activities in the logistics chain ensue in most enterprises in the scope of IT-based production planning and control systems. Theses systems are also called PPS-Systems.

1.2.3   Procurement logistics

A high and flexible reaction capability towards the customer requirements depends to a great deal on the supply with input goods of external vendors. Here procurement tasks have to be used. The procurement logistics has to ensure, that the industrial enterprises receive sufficient raw-process materials and operating supplies, which are necessary for the production in time and that the trading enterprises can provide the products, demanded from potential customers, as fast as possible with the minimization of the stock and transport costs. Following tasks are included.

- Requirement determination and disposition

- Supervision of deadline of delivery and delivery amounts

- Supervision of quality-, packing-, transport- and shipment specifications

- Goods receipt and control

- Stock and inventory administration

- Operational transport

- Planning, control and regulation of the material and information flow

1.3   Information technological realization

Today's requirements of the market force the enterprises to permanently improve the quality and performance size of their products. The most efficient method hereby is, following the experience of the praxis, to consider and optimize structures and business processes. The slogan Business Reengineering is in every mouth. Business Reengineering and the attached concepts and methods have in meantime grown to a paragon of programs, which -on grounds of far reaching structural changes- require a fundamental renewal of today's business and organization structure. With consideration and radical reformation of central business processes, the Business Reengineering can lead to significant improvements.

Every enterprise therefore tries hard to perform it's own organization form, the production running and certainly the products, according to the markets requirements. The successful company of the future will therefore extremely differ from today's enterprises. These differentiations will certainly affect the whole area of information technology.

Information technology is regarded as enabler for process orientated reorganization measures and is an important success factor of the business reengineering. The information technology is on the one hand an important factor with the procedure of process orientated reorganization measures. On the other hand the change of business processes leads to a change of the corresponding information technological support.

Since fact is: The one thing, which is not affected by the processes of change in an enterprise, is the fact, that everything permanently changes. The resulting conclusion can only be the following: An enterprise must slow high reaction capability towards the market requirement, and continually adjust, to the corresponding requirement. With that the enterprises also make most of all clear demands on the information technology. On the one hand important information have to be at once central and use overlapping available, that means, in every time, from every place, the access to these data in real time has to be possible. It's not worthwhile if the data don't reflect the present situation, but that of the week before. Real processes, which up to now followed each other chronologically, must now take place parallel. In order to receive maximal productivity, it is necessary, that the corresponding data are automatically available in the various deployment areas, this means that they're deposited in a central database. Data redundancy is therefore excluded and possible errors are avoided. Further more requirements on a future fitted ERP-systems (Enterprise Resource Planning-System) demand that all business processes of an enterprise can be represented process orientated and structured.

The SAP AG, with it's headquarter in Walldorf (Germany), offers with the operational standard software R/3 a world wide available operational standard software system, which fulfils all described requirements in an eminent way.

The R/3 software is a multistage client-server system with an extensive operational functionality, which is distinguished by enterprise wide integration. Thereby united data access, flexibility and productivity on highest level are guaranteed.

With the help of these software packages, enterprises of the various branches as well as public administrations, institutions, hospitals etc. can continually adjust to the permanently changing operational requirements of the market and therefore strengthen their competition position.

In the following chapter the R/3 software is to be looked at in detail.

2 Implementation of standard software

2.1 Situation

The business strategic, economic and inner organizational conditions in enterprises -this is also referred to middle class- require in the scope of a business reengineering, the implementation of operational standard software

2.2 Definition

"The rise of standard software continually increases in today's enterprises of all branches and magnitudes." [Lehner94]. Referring Stahlknecht, standard software is a finished program package, which consists of a variety of programs and together covers a close enterprise application field. These program packages have been developed in such a way, that they satisfy a greater number of users for their types of problems [Thome 96].

This results, referring to Thome, in eminent cost advantages and a higher quality control. This is to a great extent, for all fields of the system software (operating systems, database administration systems and so on) [Thome 90].

2.3 Attributes

Referring Stahlknecht, the program packages show following attributes:

"The program packages undertake an exactly defined operating application field, e.g. invoicing, assets calculation or personnel accountancy. The program package or the single programs have a fixed price for the basic version. The adjustment on individual operational requirements is calculated and fulfilled." [Stahlknecht 95].

Standard software is according to definition, mostly sector independent [Stahlknecht 95]. Even user-orientated programs can occur in form of standardized software, but in this case they are mostly called standardized software user software. User software, which is created to the requirements of certain branches, is called 'branch-software' [Stahlknecht 95]. In this context it can also be spoken of so called 'branch models' (also called 'pro configured systems') since there are no fundamental differences with the characteristics and the selection process between branch and standard software, one generally speaks of standardized software. Standardized software for branch neutral use is most of all offered in form of modularly built-up, integrated packages, whereby the single modules correspond to the operational fields of work (accountancy, logistics and human resources) [Stahlknecht 95].

2.4   Customizing

The programs must still be adjusted to the special requirements of an enterprise [Stahlknecht 95]. For the adjustment of standardized software to the operational requirements, also called 'Customizing' (= adjustment of a standardized software to the special requirements of the customer) [Thome96] there are three possibilities:

- Parameter settings
- Configuration
- Individual development

2.4.1   Parameter settings

Here the setting of the parameters initializes the wished functions of the program. An essential prerequisite is that all possible program functions are available.

2.4.2   Configuration

Only the program components needed are selected and integrated within the software package during the configuration. The invoicing takes place by PC with the selection of the existing components.

2.4.3   Individual development

This means, that the software for the necessary implementation is individually provided. Such modified software masters customer requirements the best, but it is also the most expensive solution.

An alternative to the adoption of standardized software is the adjustment organization process e.g. by reorganization of business process, which can also cause a change in the build up process. The synchronic adjustment both of the standardized software and the operational organization is often the most suitable solution [Stahlknecht 95].

2.5   Implementation success factors

Jakob/Uhink mark the implementation and use of operational standardized user software by following success factors.

- Harmonizing of business and information strategies
- Fast realization of run able solutions
- Concentration on 'early wins'

In the following these success factors shall be described in detail.

2.5.1   Harmonizing of business and information strategies

The implementation of standardized software must follow the business targets of an enterprise. Decentralization and segmentation of enterprises parts, reengineering of basic business processes, business process optimization, increasing of the productivity of organization components must be supported by an corresponding user software. But in most cases modern standardized user software as SAP/R3 enables a new business orientation in sense of the target mentioned. For a lot of enterprises the use of standardized user software becomes a strategic decision.

2.5.2   Fast realization of running solutions

An essential success of the implementation of standard user software is, to faster gain running, but up to now not already full detailed solutions. Therefore it is possible, that all employees can at once have a clear idea of the new business processes. Declared and primary target must be, to accelerate the implementation of a standard user software package by fast and efficient realization of successful basic solutions and the repeated further development of the basis solution with an iterative prototyping.

2.5.3   **Concentration on "Early wins"**

In addition to the already mentioned success factors, projects must be carried out and soon lead to a measurable benefit. With that the project team should first of all focus on the so-called 'early wins' before comprehensive business process optimizations or reengineering measures are started. "Early wins" means the improvement of business processes, which are yet realizable a fast way and walk along eminent costs and/or time saving effects.

2.5.4   **Laying upon standards**

An essential success criterion for the determination of the support and maintenance costs of the standard application software is the declared preparedness of the enterprise or the project team, to use the software, if possible, in standard and refrain from expansions or modifications. Modification requirements often result of the demands of the users, that the new software shall correspond in handling and dialogue conduct to the former used system [Jakob/Uhink 98/].

2.6   **Targets of the use of standard applications**

"Essential targets of the use of standard applications is, the reduction of customer specific software development, training and servicing costs. Further positive effects, which are expected of the use of integrated standard applications, are:

- The maintenance of the integrity of business data

- The immediate availability of all business data for every employee

- The field encroaching view of operational processes and objects as e.g. order, booking/reservations, invoice etc. [Keller 97]

- A flexible change of operational processes within the solution spectrum offered by the software provider"[Keller97]

2.7   **SAP – The company and its successful software product**

Dietmar Hopp, Hasso Plattner, Klaus Tschira and two further former IBM programmers have together founded the software house SAP GmbH with headquarter in Walldorf – Germany. The letters SAP are the abbreviation of Systems, Applications and Products in the data process.

Target of the enterprise was and is, the development and purchase of one single operational business standard application, which covers all operational fields and offers a homogenous structure and user interface.

2.7.1   The product

SAP R/3 is an integrated vertical neutral standard application, which covers, iterates and connects the fields accounting, logistics and human resources. The SAP Software covers the whole operational user spectrum in industry trade and service as well as public administration. On basis of SAP's own program language ABAP/4 a particular development surrounding is provided. The consequent integration and the extensive setting of parameters, which permits a flexible adjustment to business specific requirements, are special characteristics of the R/3 software. The comprehensive orientation of SAP-systems supports most of all the optimization of business processes. The "R" of R/2 and R/3 stands for real-time. And emphasizes the immediate transaction and actualization of data, which are available to all departments affected by this integration. This is very important for an enterprise on grounds of competition, since the data must project the present level and not that one of the week before. In other words: No quantity transaction takes place without the corresponding value transaction. The SAP AG offers the product in two systems: R/2 + R/3. Both are marked by following advantages:

* Comprehensive operational functions
* High integration depth
* Modular organization
* Vertical neutrality
* Clear structure
* International application [Wenzel 96/6]

System R/2

1978 the system R/2 was implemented, which is even in present times used in a lot of companies. The R/2 system corresponds to the requirements of companies with mainframe orientated structures. Target groups of the R/2 system were therefore groups and big industries. Actual over 1600 R/2-Installation do exist world-wide.

System R/3

1993 the breakthrough of the SAP AG came with the new so called client-server system R/3. R/3 is with its open conception orientated to the use of client-server architecture, whereby the applications can be distributed on several server-layers. The program allows a more flexible use of business data to the companies. Further more the R/3 software is designed for the use of various operating systems. With that R/3 produces a technological revolution. Since the SAP AG doesn't merely stand on the first place of the German software house, but also belongs internationally to the top flight.

At present R/3 is worldwide the standard for operational application software. The system R/3 has in medium-term, to be regarded as the successor of R/2. Surveys on customers of SAP AG attest, that nearly all present R/2 users plan a migration to the R/3 system or even have begun. This however takes place most of all in a coexistence to R/3.

The R/3 System includes following specifications:

Functionality

The R/3 system is marked by an all-embracing functionality. Finance e.g. profits from the functions of the application areas as purchase, sales or human resources, from which it achieves the transaction data. The integration of application guarantees the availability of all functions in the system and with that in the company also.

Internationality

With R/3 enterprises, internationally operating as well as multinational companies, can carry out the business processes of different sites on a joint server and deal with country overlapping procedures in one system.

Vertical neutrality and adaptability

With the so-called 'Customizing' as well as the making out of possible separate modules 'vertical solutions', the adaptation on the R/3 system on vertical specific characteristics, take place. The vertical neutrality of the R/3 software becomes clearly apparent by taking into account the most different branches. The R/3 system can be applied both in large groups and in middle-class companies.

Data base consistency of R/3

Data are consequently stored only once in the R/3 system. Therefore these can be consistently loaded down for every evaluation in every enterprise field. Herewith the risk of unadjusted and not topical data sets, as well as constructive and therefore faulty evaluation is prevented.

User interface

The user interface of the R/3 system mainly corresponds to that of MS-Windows. This leads, compared with the R/2 system, which was not provided with such an interface, to a considerable higher user comfort. The importance of graphical user interfaces is often underestimated. Graphical interfaces make the presentation of a lot more information possible on screen. Further more, information can be received faster and more complex.

Modular setting-up

The R/3 system is divided in individual modules. Any of these modules consists on the other hand of single components and subcomponents. The different R/3-modules can be individually achieved, though it has to be taken into account that certain components are not running on their own.

2.8   Implementation of a standard user application - procedure model

During the implementation of standard application software normally a very long and hard way has to be gone, until the various possibilities of the software and the companies' demands are well suited. Only then the system can be productive [Jäger 93]. Companies, which are newly confronted with the implementation of software systems, therefore need a special project management.

SAP has developed with the R/3 procedure model an implementation methodology for SAP products, based on an experience of more than 1000 R/2-implementations. This model is supported by a function catalogue (customizing-implementation guide) existing in the R/3 system, and subdivides the IT-projects in several project phases and operation steps. Essentially, the implementation phases "organization and design", " detailing and implementation", "preparation for production" and "productive operation" can be differentiated.

2.8.1   Organization and design

In the first place, it is the organization and design phase, general preparations are carried out and a practicable implementation concept is worked out. This includes the requirement analysis, the initial entry of inventory data of the processes and functions to be applied in the R/3 system, project planning and organization, the rough design, the system installation, the model of interfaces and system extensions and trainings for the project members. Based on the R/3 functionality and the company specific requirements, a technical-and data process concept is committed. The result of this first phase therefore describes, which components and processes of the R/3 system are to be implemented and who is responsible for the realization of the partial projects.

2.8.2   Detailing and implementation

In the second phase, the detailing and implementation, the single application modules (e.g. materials management, sales and distribution, accounting) are made a prototype corresponding the former worked out concept. While in the organization and conception phase hardly company specific adjustments were implemented, the emphasis here lies on the customizing in the narrow sense. Here, different working packages are implemented. Any of these working packages can be tested for itself that means module referred. Eventually the so-called final inspection takes place. With it, the participation of all components has to be paid attention, in order to be able to test the integration capability and to provide a regular process of business transactions. After the termination of this phase the detailing and implementation of the software concept and its realization by customizing applications in the system is performed. Within the formal termination of this phase, the quality test takes place. Here in particular the project targets and test results are examined.

2.8.3   Preparation for production

In the following production preparation the, for the productive start, necessary measures as documentations, system tests and user trainings are gone through. The end result of this phase is an established productive system. In general, a so-called double installation exists in an implementation phase. This means, that a test system as well as a productive system are available. In this context the R/3 system, in which the basic data of the field-departments are processed and stored, is the productive system. Whereas in the test system new parameter settings, personally developed and changed transactions, effects of customizing changes, new puts and personal add-ons are tested. With the use of a test system as well as of a productive system the authorization and authorization profiles, reports and old-data are transferred to the productive system. The result of the productive preparation is a checked, productive system.

2.8.4.   Productive Operation

Now the phase productive operation gets started. After the data migration, activating programs can be started. The master data is still accompanied after the production start for a certain time of the project team. This optimizes the R/3 system. This procedure module for the implementation of the R/3 system has been supplemented by the SAP implementation methodology, which is described in the further course of the book.

2.9   Benefits efforts of practice by implementation of SAP R/3

With the implementation of the operational standard software SAP R3 following benefit effects, in regard of various experiences, can be achieved.

- Adjustments on process organizational changes can easier be represented

- The competent and department overlapping application of operation processes increases the security of the business processes

- Qualitative improvements of internal transactions and decisions e.g. by integrated, daily "planning runs" of customer orders, orders, invoice…

- SAP as an open system offers a variety of standard interfaces and as a consequence enables the easy connection without a loss of integration advantages to the most different satellite systems

- Due to a very transparent data management and a programming, easy to be learned, with SAP developed program language ABAP/4, the flexibility is increased

- Further benefit potentials become apparent, because of the possibility of an simple incorporation of MS office and office communication solutions

- The integration of the processes leads to a faster procedure of business processes. The R/3 document flow illustrates, which employee is included in the process chain. Herewith any one at any position of the business process is enabled to determine the status of a business process, to control and finally to recognize, if something's wrong

- Thus, not only the operation speed is increased but also the quality and security of the business process. Short delivery times and a smaller amount of complaints are examples for these positive effects

- A change of view on part of all employees of a company is required. If R/3 is implemented, the company so to speak gains new processing grounds. Have the employees in the past been used to working in a fixed procedure and fulfilling operational tasks in successive order, everything works in real-time, besides some exceptions (e.g. periodical jobs) with the R/3 system.

- This implies more responsibility and certainly more initiative of every single employee, since any input in the R/3 system immediately produces a chain of reactions. If merely one link in this chain doesn't work properly, the entire procedure is severely disturbed

- The employees even receive a better understanding for 'their' company. For the growing responsibility also forces the employees to look more and more beyond their own functions into others. The result is that the employees release their view from their so far isolated procedures and become aware of the importance of their contribution to the entire process

- The costs for the system maintenance are reduced, since the high integration of the R/3 system leads to an omission of foreign systems and also of the data transactions via "work- and maintenance" intensive interfaces to these third-party systems

- These are also immense advantages in the so-called data management. Now it is possible to offer information at any time rapidly and user friendly

- The administration expense can be reduced. Before the implementation of R/3 all customer papers have been filed several times and manually administrated. Now this is undertaken by the R/3 system. With that every department of an enterprise can again concentrate on it's main business

- The access to internet and thus to the communication technologics of the future is possible

- The system guarantees actuality and flexible adjustments on the cvaluations on branchcs

- The system delivers an improved decision basis for the management of a company by regulation and control of management ratios

- A great advantage of the R/3 use is caused by the revision of business processes in the scope of the project. Therefore it is possible to tighten all processes and to reduce the number of manual operation steps

2.10.   Consequences of the implementation

2.10.1   Organizational effects

The most essential and considerable effect of the implementation of process orientated standard software is the causal connection with the integration of tasks and functions along the different business process chains of the single departments of an enterprise. Herewith an enormous mental reorientation of the entire staff is necessary.

- In many functions and fields of duty there's an extreme high force, to think and act in a department field overlapping way to avoid a disturbing influence on the staff

- Responsibilities must be newly arranged since many functions gain in higher significance

- It is not any longer possible to perform process improperly and provisionally like before the implementation of SAP R/3. The standardized process orientated procedures require the discipline of all employees. Eventually this leads to an effective improvement of the entire data and procedure quality of the software system implemented

- A large-scale change of mind is also required from the management since later conversions in the setting up respectively procedure organization are not any longer possible without taking the SAP surrounding into consideration. Talking turkey this means that every single change even in SAP must be deposited and the resulting effects (must be) tested field overlapping.

2.10.2 Implementation consequences

The implementation of operational management standard software SAP R/3 also has personnel consequences. These are according to Uhink/Jakob:

OPERATING AND SYSTEM ADMINISTRATION   Tendency:↓

Due to the homogeneous DV-Surrounding with the most of all consecutive use of Windows NT, the system administration can be strongly reduced.

SUPPORT   Tendency:↑

The implementation of SAP R/3 requires a much higher expenditure for the support. This is based upon the grown use of pc-technology and the increased application employment of modern communication technologies as e.g. an Internet "connection" to the existing SAP-system.

SAP-USER DEVELOPMENT   Tendency: ↑

These expenses will also tendentiously increase. The extent of cost increase is defined by the necessary maintenance requirement, further development and service demand of the SAP Software.

2.10.3 Further critical factors

- The temporal expenditure for document adjustments, the preparation of users, the integration of processes and for the transmission of master- and transient data is under estimated.

- The members of the project team for the implementation of the SAP software are overburdened. On the one hand they must deal with the common and daily work and on the other hand they must additionally fulfill work-intensive project tasks.

- Too less importance is attached to company specific adjusted authorization concepts.

- The employees are not sufficiently enough instructed in the newly applied procedures. To mention are e.g. availability control, disposition procedures, and so forth.

- Transaction-, module- and integration tests are not conducted or much too late.

- In the scope of the customizing the business structure aimed was inaccurately represented in the system. If here the course is falsely set, enormous costs arise later, in order to correct this.

2.11   Recommendations for a successful implementation of SAP R/3

- With the selection of the external consulting partner, his functional and social performance capacity is to be analyzed. Thus the good or bad quality of the consulting is essentially responsible for the success or fail of an implementation project.

- If possible, one should always cling to the SAP standard. Herewith, great advantages for the implementation project can be achieved, since the SAP system has already been delivered with a variety of applications to the customers.

- The 80 to 20 percentage rule should be considered. First one concentrates on 80% of the implementation requirements, and thereby accelerates the implementation time. The remaining 20% are then realized after the productive start. The implementation methodology ASAP, developed by SAP, works essentially on these rules.

- Particular value should be set upon an absolute detailed and accurate service of the master data in all areas. It is important, to manifest responsibilities as soon as possible.

- In order to always give every single employee an actual overview over the present status of the project, they must be informed in detail but also continually and fully trained. This obviously increases the innovation of the employees bound in the project, since they learn to better estimate their part in the project success.

- Great value must be set upon professional project management.

- It is advisable, to take time for a detailed and comprehensive integration test, during which not only the functional consecution of the system are tested/analyzed, but also its load.

- From the start the employees must be intensely involved in the project. Further more, high importance must be set upon the training of the users. The quality of the training measures increases the more, the better the training is tailored to the particular needs of the employees to be trained

- The source data take over requires an accurate planning, organization and realization. In this connection it may be referred to the book "Testing SAP R/3 Systems" by Oberniedermaier/Geiß, published in the SAP Series of Addison Wesley. It describes in detail a very comfortable automatic procedure for the takeover of source data with the SAP tool CATT- Computer Aided Test Tool. This procedure has already stood the test in proactive and can therefore nothing but recommended.

- Detailed and increasing documentation of the project progress, for the new employees can any time get acquainted with the current work and duplicate the project step by step. With that, a dependence both on external and also internal project employees can be reduced. If an internal know-how owner quits the company, the documentation and also the knowledge of the project stays in the enterprise, at least.

- Concentration on the essential basic processes in the company. This guarantees a clearly arranged, simple and with that also a risk less conversion.

3 Logistics with SAP R/3

The operational logistics, as already described in chapter 1 in detail, comprises the entire performance of material, information and production flow from the deliverer over the production to the consumer. The user modules of logistics offered by SAP make it possible, to plan logistics procedures of all sorts beyond field borders, to control and coordinate. This becomes possible among others by the data and function integration of the SAP software.

SAP R/3 is equipped with the following user modules of logistics:

SD : Sales and Distribution

MM : Materials management

PP : Production Planning

QM : Quality Management

PS : Project System

SM : Service Management

PM : Plant Maintenance

In the following the single user modules of the logistics are described.

3.1 Materials management

The module MM – Materials Management enables the company to handle the entire process chain of the procurement cycle proceeding from procurement demands to the goods receipt.

The procurement of raw materials and the material flow with a company play an essential role for the success of an enterprise. The use of materials management of the R/3 System includes all necessary functions in order to simplify the processes of stock planning, purchase, stock control as well as operational accounting and invoice verification- standard procedures. These functions are connected with each other and integrated with the remaining functions of the R/3 system. With that the materials management and the other users of logistics and operational accounting are provided with actual information at any time. With the easing of routine processes, time for more important tasks is gained.

Aside from efficient procurement processes for materials, the application MM offers powerful productive tools for a rapid and inexpensive purchase, the taking and the temporal account of services.

For negotiations with deliverers, the purchase information system and the deliverer evaluation provide the information necessary. It is possible, to define for ones own, which data have to be taken into consideration in the survey and how this information can be prepared. With the stock controlling for example stock items or transshipment frequency can be considered and analyzed. These information systems offer perceptions about appearing deliveries and therefore supply the grounds for the decisions necessary.

3.2 Production planning

This module deals with the quantitative and temporary planning of products to be produced and with the control of the production processes itself. In addition to the master data maintenance even all quantitative and capacitive steps for the planning and controlling of the products are supported. Examples for similar planning concepts are e.g. MRP II or KANBAN.

The application PP includes the complete manufacturing process from the registration for the production master data, over the production planning, material demand planning and capacity planning, to the production control. It is designed field-

43

overlapping and offers a variety of production methods, which lead from the single or variant production to the serial respectively mass production. Like the other R/3 logistics applications, the application PP is also fully integrated into the entire R/3 system. Independent from, whether a single system or a world wide net of distributed R/3 systems are required, which are connected by ALE with each other, the data are not only immediately available to the production planning and control, but also to the accounting, sales, human resources.

3.3   Quality management

This module is engaged with the scope of duty of guaranteeing along the logistical process chain, that the processes as well as the units prepared in the processes can meet the requirements laid upon them.

With the application QM, the procedures for the achievement of a high quality standard in every phase of the logistical chain, can be systematically realized. With regard to this target, the application QM is narrowly integrated with other applications in the R/3 System. The tying up in a logistical system offers advantages, which are only reached by a decentralized "island solution" with great effort.

The purchase receives current quality ratios for the deliverer evaluation as well as quality-referred data for inquires and orders. With critical materials, a vendor release can especially follow on grounds of the quality system. Control data, given by the quality management, lay down which materials are to be checked and laid into the quality test facility – for example because of goods receipt, production- or acceptance control. With that it is guaranteed, that merely those goods are transferred to processes, which meet the given quality requirements.

3.4   Plant maintenance

All performances connected with the planning and procedures of repair work are supported by the module Plant Maintenance. This includes the termination of periodical servicing and inspection measures as well as the placing of orders for the own and foreign orientated maintenance of unforeseen disturbances. Not only company internal as e.g. production, supply-, disposal- and transport systems but also systems as e.g. customer sites can be administrated by the maintenance module.

The application PM offers technical and accounting evaluation, whose performance can vary by the most different criteria (e.g. organization unit, site, implementation period of measures or producer of the installation). These information help to minimize the period of defect referred installation lay-offs, to reduce the resulting costs and to recognize in time possible weak points of operational/business installation. They form the basis for the investigation of an optimal maintenance strategy in the sense of a TPM or a risk orientated maintenance. The interface free integration of the maintenance in the R/3 system is the basis of a rapid and efficient communication and cooperation of all business fields. With that a complete vertical neutral maintenance management system is available, which understands the maintenance as an integral part of the enterprise resource planning. In many approved practice employments the application PM has fulfilled the requirements of the technical and commercial business fields to a complete installation information system.

3.5   Project system

With the help of this module, projects of all sorts can be represented and administrated. Special features of projects can be taken into consideration. In general, projects define themselves as complex, temporal limited, cost- and capacity intensive projects. They are built on exact target terms and defined quality requirements. Because of these prerequisites the projects have in general a high strategic meaning.

The project system analyzes and controls the availability of financial funds, capacities or materials so that they really are sufficiently available, if needed for the project realization. The project costs can be limited and controlled with the help of authorized and released project budgets. SAP Business Workflow improves the communication and information flow in huge projects. Thus, the purchase for example is immediately and automatically informed about amount or deadline changes.

During the entire project procedure, an efficient and flexible information system is available, with which layout and specification levels can be customer orientated defined. Evaluation in list form or graphical analyze for budget, actual and budget costs, earnings, obligo deposits and discharges, deadlines and resources give all necessary and demanded information. Standardized interfaces offer the communication basis for a common data exchange for further planning and calculation in additional systems.

3.6   Service management

The increasing globalization of markets leads to an ever-hardening competition between the companies. After a change in the business targets, today's service management plays an important role as an independent business branch. The modern system landscape is marked by a missing interpretation between the financial systems (e.g. service, administration, invoicing) and the rather technical systems as customer object administration, service procedure and so on. It is rather impossible, to rapidly make decisions on basis of service contracts obligation and resource availability. With the service management system a highly integrated, autonomous vertical solution is available.

The administration of the customer's service objects with configuration and history is a central part of the system.

Long lasting service agreements are fixed in service contracts respectively guarantee conditions, which are analyzed with the record of a service call. With the service procedure the recorded customer calls are automatically examined about service contracts obligation and guarantee conditions. For the registration of actions, damage symptoms, damage cause, spare part finding and so on even free data formats can be used aside of a comprehensive feature catalog. All service costs are stored in a history database. More over an efficient information system is available, which can be adjusted according to the demands of the enterprise. Clearing and invoicing can ensure within service contracts periodically and within the service order according to time and material expenditure.

3.7   Sales and distribution

Today's market powers force enterprises, to simultaneously improve their customer service, to eliminate the total product cycle from order over delivery to invoice office to improve the quality of the sales net and reduce the sales costs. There are three methods for reaching those targets: Business optimization, the employment of new technologies and the strategic cooperation with customers.

The demand on strategic cooperation is referred to the fact, that the market tendencies are increasingly determined by the customers. Customers expect, that their needs are met to face the hard competition about customers. Far-sighted companies develop close business relationships to deliverers and customers. With that, they minimize competition while they maximize their possibilities of recognizing the customer requirements and accomplishing. With the modules of the R/3 system any of those methods can be successfully applied.

Due to the sales module all those exercises are supported, which are necessary for he realization of activities within the sales logistics. With the sales module products can be sold to business partners and even services produced. All necessary data are filed in so called master data. These master data form the basis for all activities in the sales transactions. With that the SD module takes charge of, besides of administration of those master data, the execution of all tasks of sales, transshipment and invoicing.

The following chapter deals with the questions of how the sales logistics with SAP R/3 can be applied in detail.

4   Information system to selected fields of sales

As already commented in the introduction chapter, all tasks, serving the organization of supply activities, belong to the sales – respectively distribution logistics. That means to hand over the customer the demanded product at a defined place, to a defined time and if wished, in a fixed combination with other products. To meet these very complex requirements, a standard application must fulfill diverse integrative tasks.

In the sales of a company those tasks are divided as follows: First the customer orders must be placed. Sales "representatives" therefore make customer and market analyses, contact the customer in writing, by telephone or personally and carryout special purchase actions. The customer can now place an order and receives a quotation. He can accept this quotation, which leads to an order. Now it must be analyzed, whether the material is available or must be ordered, respectively packed and delivered. Eventually an invoice must be prepared for the customer.

At first sight, most of the software systems regarding their functionality, meet this very superficial description of the sales process. The differences between the systems lie in the detail and become very fast apparent. Many systems fail, if e.g. a third party deal or a customer single order is to be applied or a system supported returnable packaging or consignment procedure is to be established. A very comprehensive software package, referring the sales functionalities and the level of integration with the single business field is the operational standard application software R/3 of SAP. The operational requirements on the system shall be defined for the different fields, and the level of fulfillment shall be represented by special functions within the system.

4.1   Application of the sales within the R/3 system

To be able to display the sales of an enterprise within R/3, different structures, data and transactions must be filed in the system. This takes place in form of following basic structures:

- Organization structures (sales external- and internal organization)

- Master data (material master, customer master, deliverer master)

- Sales documents (purchase-, shipment- and invoice documents)

The structure of the particular company is performed by the organization. Here, the user finds numerous organization elements, which help to present his individual sales organization structures.

The data for materials, customers and deliverers are filed in master data. Different sights are implemented for the various company fields, since e.g. on the material master the sales as well as the production or the material economy have access.

The actual business transactions in the SAP-system are handled with the help of documents. Documents include important control elements referring the procedure and the structure of single sales transactions.

4.2   Survey of efficiency features of the R/3 module SD

With an efficient sales procedure, integrated processes combine all activities necessary for the fulfillment of customer requirements with each other. With a customer order the ordered products therefore pass through the entire logistics chain with the maxim of highest customer satisfaction with the most possible low stock.

The system SD enables the quick and aimed implementation of sales requirements of an enterprise and the worldwide supply of data and information. The system SD is a process controlled, as a building stone in the R/3 system integrated, application module.

The system SD presents following efficiency features of the sales procedure:

- Smoothing, efficient workflow because of comfortable business specific sorts of transactions and functions

- Data exchange both within the system SD and between the particular application, as well as quick operation of transactions and data consistence with the help of integrated transactions

- Registration and less sources of error by automatic proposal of data from the master data

- Immediate application of the current status of a business transaction in the system

- Stronger distribution possibilities on the international market on grounds of foreign trade functions

- Smoothing communication because of multilingual and immediate available documents.

According the operational requirements it should be displayed, to what extent the offered functionality are sufficient respectively usable and in what field are deficits. First of all this is described for the field organization and then for the special fields purchase, shipment, invoicing and the sales information system.

Multilingual multi-currency procedure

The multilingual order preparation with different currencies for many companies enables the single registration of an order for a limitation free process in one language. Processes with partners in other countries can be automatically translated in other languages by SD. The partners of a company receive the single process in the corresponding language and currency. Due to the simple procedure of these limitations free business processes the further development of the global market is supported.

4.2.1   Organization

The sales organization of a company is realized on the first hand extraordinarily by functions, on the other hand also by products, consumer groups or purchase fields [Witt 94]. The level of structure in different organization elements depends on the diversification of products, customers or of special conditions.

Every application of the R/3 System has a personal organization unit. Thus, personal organization units of the sales can be defined in the system SD, which differ e.g. from the ones of the materials management. Because of that the single business fields can define special meaningful organization structures and terms (which are meaningful especially for their fields).

The R/3 system enables the integration of the entire business structure without the loss of control over single business fields. With help of different organization levels in the SD business fields can be defined by differing point of views in order to efficiently administrate and control business transactions. Simultaneously a proper information flow results from the integration of single processes beyond all organization levels.

SAP R/3 here offers the possibility of defining the elements sales organization, sales paths and fields for the sales external organization. For the structure of the internal organization sales offices, customer groups and salesmen can be filed.

The shipment can be structured by shipment- and loading offices, which are related to a plant. All common sales and shipment structures as e.g. central, customer- or material referred organization can be represented in the system by different relation possibilities with these organization elements.

4.2.2   Sales

Information systems in the field sales should fulfill following tasks and functions:

- Support of the marketing e.g. by planning of marketing actions or registration of evaluations
- Planning of a sales and production program
- Order preparation/process (from the order receipt over the shipment to invoicing and the transaction to the financial accounting) [Scheer 90]

The requirements raised by MERTENS upon the functions of an information system for the sales, are similarly formed. These are described as follows:

- Control of the salesmen employment by transaction: selected customers, travel route and product offer
- Inquiry and quotation process

- Quotation control

- Quotation registration and control (technical test, solvency and rating). With the routine quotation registration one screen is sufficient, in order to register customer, products and amounts. The other data are taken from the master data and proposed.

Depending on the application of the availability examination and the credit check, the quotation can be booked with only one mouse click.

Efficiency feature of sales procedures with SAP R/3:

- Comfortable quotation registration and customizing possibilities

- Automatic determination and flexible definition of prices, discounts and surcharges as well as taxes for a business transaction

- Reduced data registration with the sales procedure by integration of the processes into offer R/3 modules

- Lesser credit risks by comprehensive functionality in the credit management

The functions offered in R/3 for the sales procedure meet these demands. The marketing and control of the salesmen employment are supported by the module CAS - Sales Support. The sales plan is enabled by the sales information system. The different functions for the quotation process are represented in the following.

Sales support

Sales methods are changing. Nowadays traditional sales methods are often replaced by business relationships, which base upon the cooperation between producer/vendor and customer. Instead of the typical short-term relationships, based upon opposite interests, the customer now looks for long-term partnerships, which are built on reliable and frame communication for common advantage. The information management has become, compared to former times, a far more important component of sales.

The sales support enables the registration and evaluation of customer contacts. Competition data can be, like market data, registered and recruited for evaluations. Tendencies/market trends can be prematurely recognized, by examining current data and comparing those with plan data. Then the necessary steps for the solution of problems can be initiated or the offered possibilities used. The most important information can be filtered out of the mass of data quick and with any specification level filtered. Moreover there is the possibility of controlling aimed marketing actions as e.g. telephone marketing, or mailing activities. But automatic finding of the travel route of key account managers, as required by MERTENS for a marketing information system, exists in R/3 only for the field shipment.

Supervision of sales transactions

The single document types control the sales transactions. With these it is determined, which additional functions as e.g. availability examination, shipment termination, price finding or route finding should be implemented. R/3 offers a very simple standard solution for various business types from mere trading companies over service companies to quotation orientated construction plants. The sales documents serve for the transmission of information and form an integrated sales information flow. The sales document data are automatically taken over to the shipment and the sales and shipment data on the other hand to the invoicing. With the help of this document flow information can be transmitted with minimal costs. As soon as the data registration and control is finished, the data get automatically transmitted.

Shipment appointment, availability inspection, demand take over

The appointment inspection, the inspection of the availability and the take over of demands to the materials management are also supported by the system. With the availability inspection it is examined if there's enough stock for the time of the planned delivery, before the termination of the quotation registration, if there is not enough stock for the immediate shipment, it is directly determined by the availability inspection with real-time processing. The transparency of the customizing suffers under the great volume of functionalities and the high level of integration. The system indeed meets here most of the requirements, but adjustments in this field are often long termed and connected with very high customer expenditure.

Pricing

The pricing also takes place system-supported. Beside various pricing conditions also discounts- and surcharges can automatically be established and if required, material changed. For the application of the pricing the condition technique is used in the R/3 system. This is because of the functionalities to be performed so complex, that alterations require an enormous high temporal expenditure as well as special qualifications of the consultants. The pricing is an example for the dilemma of the system developers, to unite as much functionalities as possible and still guarantee flexibility referring the individual adjustments. The flexibility and the comprehensive functionality of the pricing are an essential power of the SD module and support even the most pretentious sectors of industry. With the help of pricing rules even complex pricing scenarios can be filed in the system. The advantage of this functionality is most of all that the employees of sales must not any longer prepare a complex pricing, but can concentrate on their main business, that is sales and services.

Credit limit check

The system SD allows a flexible credit check. For every customer an individual credit limit can be declared. R/3 then conducts an automatic credit limit check within the quotation registration and the resulting reaction follows. Credit limit checks can be made at different times within the sales cycle, from quotation receipt to delivery or but from single or groups of comparing branches in own responsibilities. It is possible to define a credit limit for a specific customer and/or specific credit limits for credit control areas. Further on different system reactions can be determined for the case of an exceeding the overdraft.

Generating of trading documents

The generating of quotations, order acknowledgements, delivery notes, picking lists, freight lists, packing list, invoices as well as credits, debits is supported by the system. All common communication media as e.g. EDI, Mail, Telefax or the simple printing can be applied, but the adjustments to be made of a formula are often very complicated.

4.2.3   Dispatch

The dispatch follows the sales procedure. Is a material free of deliverance, it must be first off all picked. Now the dispatch type, the delivery store and the transportation type are to be determined, as well as the transportation route fixed. Further on the dispatch documents should be automatically prepared. In an integrative system the data necessary for the delivery are drawn from the order. This base function should any integrated application system fulfill in the area of sales.

Following functions of the sales are to be performed referring MERTENS:

- Picking

- Delivery release

- Dispatch logistics (determination of delivery store, transport type, load, travel route, optimization possibilities and preparation of dispatch documents)

During the dispatch and transport handling in the R/3 System it is guaranteed because of the narrow integration of the systems sales, materials management and financial accounting that the dispatch data is any time available in the system. The dispatch is an essential part of the logistics chain, whereby the main tasks lie in the ensuring of the customer service and in the support of the distribution planning. With help of the flexible SD dispatch handling the general cost efficiency and the competition capability can be increased. The requirements at a system supported dispatch handling are covered in R/3 in following areas:

Generating and processing of deliveries

Not only single deliveries of an order can be delivered together but also orders based upon different criteria. The integration to sales and invoicing is given and the data important for the delivery are automatically overtaken respectively transferred. A manual adjustment of these data is possible.

Picking and packing

The picking store lets automatically determine any deliver position. More far reaching systems cannot only determine the good to be delivered but also the suitable/proper packing [MERTENS 93]. This is also possible in the R/3 system in regard of different criteria. Different packaging sources can be filed and hierarchically structured. Moreover there is the possibility of automatically generating and printing packing lists. With the term sales logistics Mertens connects the determination of the delivery store, the selection of transport type as well as the determination of the load and the travel route. To this serves the automatic dispatch center finding the possibility of the consideration of weights, volumes and the route finding.

Dispatch center finding

The dispatch center can be automatically determined referring the order position in dependence of the dispatch conditions of the ordering party, the load group of the material and the delivering plant.

Route finding and transport determination

With the route finding the optimal transport route, the means of transport and the transit time are determined. Moreover the customs offices can be proposed for the export. The route finding is a very comprehensive function, whereas its high maintenance efforts are in no proportion to its results. Most of all in middle class companies the manual of means of transport and the selection of the transport route will often show equal results.

Should a company nevertheless decide on the use of all functionalities of the route planning, this will still show up some defects. The residence periods at a customer, which were important for an exact transport determination with several knots, are not taken into consideration. This counts further more for a limitation of the total transport duration like e.g. required for the transport of food or chemicals.

The requirements on an automatic, integrated dispatch process is in all areas fulfilled by SAP.

The following efficiency of dispatch processes in R/3 shall be mentioned:

- Adjustment of dispatch processes on company specific requirements by flexible delivery handling

- Picking-, packing-, loading- and transport functionalities for total dispatch solutions

- Guarantee of deadline reliability by supervisions of delivery deadlines

- Efficient information and goods flow by flexible communication

- Reduced data registration and common availability of current information by narrow integration with the systems MM and FI

4.2.4   Invoice

The last link in the SD process is the invoice and the transmission to the financial accounting. A system supported invoice in an integrated system should be able to automatically generate invoices from the order and dispatch data as well as from the customer and part master data. Herewith the possibility should be given, to determine surcharges and demands on different criteria by the system [MERTENS 93]. Further more, the handling of special invoices as e.g. invoices, which should be handed out before the delivery notes, must be possible. The system should automatically recognize those with changing consignees and cost receivers.

The system R/3 fulfills in the area of invoice following functions:

*   Invoice generating on grounds of deliveries and costs

*   Credit- and debit memos generating on grounds of requirements

*   Reversals

*   Transfer to the module Financial Accounting

Invoice generating

Separate invoices for each delivery as well as collective invoices, which are combined on different criteria, can be generated. The transmission of invoices to the financial accounting is also processed for each invoice or for an invoice list. Collective invoices, invoices lists as well as the invoice split can be defined in structure according to customers or transactions.

Processing of complaints

In the processing of complaints credit and debit memos are assigned. Should special credit memos as e.g. credit memos for empties or mere value credit memos be performed, these are very easy to define as new invoice types. Moreover there is the possibility of working with blocks in these areas, in order to gain additional control over the allocation of credit memos. With returns-, credit and debit memo processes in the system SAP, those transactions can be handled, in which the customer sends back goods. The complaint functions include deliveries, returns, credit memos, debit memos free of charge with or without reference to preceding sales transactions. The system supports the proper termination of these transactions, by proposing a delivery and invoice block, if the transaction is inspected by another department. Summing up, there are five procedures in the processing of complaint:

- In a delivery free of charge a good schedule free of charge is sent to the customer

- A subsequent delivery free of charge replaces missing or defect goods bought by the customer

- With a return defect goods, sent to the customer, or goods, which have been delivered on approval, are handled

- A credit memo demand relieves the customer account in case of amounts too high or returned goods

- A debit memo demand charges the customer account in case of amounts calculated too low

Reversal

There is the possibility of reversing invoices and credit memos. The system generates for that a reversal invoice with reference to the original invoice. The transmission of documents necessary for the reversal to the financial accounts takes place automatically.

Integration of the materials management

Due to the integration of MM an order requirement in the purchasing is automatically generated during the filling of a customer order. Then the ordered materials can be directly sent to the customer or but transmitted to the warehouse and brought together with other materials in a customer order.

Connection to module FI financial accounting

The particular invoices can be directly transmitted to the financial accounting. On demand, even here blocks can be defined, which stop the release and make further handling of the invoices possible. For the registration of the revenue accounts, which are affected by the transmission of the invoice data to the financial accounting, serves the revenue account finding. That can also be adjusted by the sales. Base for this is the condition technique, known from the pricing. In contrast to the pricing however, it is possible to make an analysis of the account finding for a business transaction, in order to reveal inconsistencies.

The functionalities of the invoicing in the R3 system also meet most of the business administration demands. Adjustments as e.g.

the supplementation of invoice types are not as complex as in the area sales.

4.2.5   ### Sales information system

New technologies are ever faster developed and their consequences far reaching. Many companies are under the pressure of recognizing a competition advantage early enough and using it. The optimization of sales belongs to the main targets of a company. This can be achieved with the help of a sales information system.

A sales information system collects information from the internal, market orientated information flows [Thome 97]. This information should be available for different individual evaluations and also serve as base for planning.

One of the most important tools of the module SD is the sales information system. With help of the generally available, current data of this information system, customers can achieve a qualitative high consultation, which on the other hand results in a competition advantage. On basis of consistent current data even efficient marketing measures can be made. All data of sales, dispatch and invoice slip in the sales support over the sales information system e.g. in form of customer master data sheets or statistics for an order receipt.

The sales information system in R/3 is part of the logistics information system LIS, to which even the production information and purchase information system belong. It enables both standard analyses and flexible analyses referring individual criteria. The corresponding data can be graphically represented and handled. Besides analyses, the sales information system in R/3 also offers the possibility of recording planning data. Those can be in table form and graphically compared with the actual data in form of evaluations. The sales information system is a tool, which is easy to handle and whose customizing effort keeps within limits despite of its volume.

4.3   ## Evaluation

The R/3 system fulfills nearly any requirements on an integrative user system for the sales. The performance of the system is a problem in the operative application e.g. in the processing of huge orders. Because of the variety of functions to be made and the complexity e.g. of the pricing, long calculation periods are necessary. This makes e.g. an order entry by telephone, which requires a quick processing time, nearly impossible. Nevertheless the problem can be reduced, by improved hardware and the reduction of necessary functions. Although the complexity of particular functions makes the adjustment of individual functions and transaction very expensive and difficult, the standard software package R/3 is nevertheless a flexible system, with which nearly any business process of sales can be performed.

Selected processes in sales

A sales process describes in general the procedure of a business transaction of a business administrative typology. Herewith a business scenario forms the model for a business process. All processes necessary for a complete handling of the tasks, are temporarily and logically connected with each other.

To meet the increasing demands, on product quality, product variety and availability and the ever-shorter lead times, the companies unite the business procedures referring customer needs. Efficient business processes are generated which help to coordinate also activities necessary in the sales, to generate customer demands and eventually also to satisfy. With the business transactions in the business process area sales of the R/3 system, sales activities can be effectively handled. The here realizable business processes support the sales, dispatch, invoice, sales support and sales information processing. Further on the sales can be integrated in the procurement and production planning, which results in a reduction of the lead time within the value added chain.

In the further course of this chapter a detailed survey of possible scenarios is given.

- Third party order processing
- Returns and reversal procedure
- Consignation processing
- Returnable packaging and empties
- Presales procedure
- Customer relationship
- Delivery of models and marketing means
- Sales from stocks to industrial consumers
- Sales from stock to internal customers
- Sales from stock by scheduling agreement
- Customer inquiry handling

- Customer quotation handling

- Customer order processing

- Free charged delivery

- Credit and debit processing

- Credit limit check

- Contract handling

- Standard order with prepayment

- Model delivery

- Tank car processing

5.1   Third party order processing

5.1.1   Definition

With a third party order the customer orders products from a company, which doesn't produce these by itself respectively doesn't have these in its warehouse. The delivery of the articles required by the customer does not take place by the company, which accepts the order. Instead of this the order is transmitted to an external deliverer, who sends the product directly to the customer. But the invoice is sent by this company, which accepted this order. The deliverer on his part invoices to the order accepting company.

The scenario of the third party processing is graphically described in the illustration 5.1.

The transmission of an order to a deliverer in general takes place if the company itself does not produce the product. Due to the transmission of an order also production bottlenecks can be equalized (no capacities, cheaper production by external deliverers, trading target and so forth).

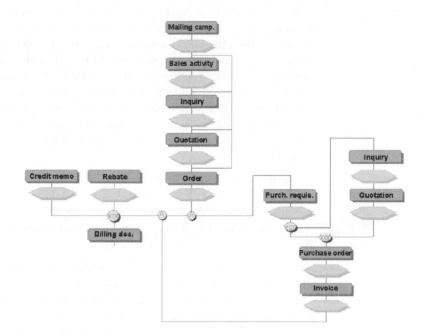

Illustration 5.1: Diagram third-party order processing

The third party processing ensues by a special order type (as far as all order positions are externally obtained)

First the registration of third party business is identical with the registration of normal standard orders. That means, the scenario for the third party order processing starts with the record of contact activities with customers. The reference to existing quotation documents is possible. A third party order can also be directly generated by a customer. Now the pricing is made and the credit check begins. With the order securing, which contains one or more third party positions, the system automatically generates a purchase requisition in the purchase. For any third party position in the customer order a matching order requirement is here automatically filed. The number of a generated order can be found in the document flow of the customer order. The responsible purchase employee inquires the order requirements in the MM and transmits it to the deliverer. A transmission can ensue in form of written documents or EDI. The deliverer gets with the order the instruction to directly deliver the ordered product to the delivery address noted in the original order. The deliverer therefore takes over the transport to the customer. With the delivery of goods in a third country a customs clearance forwarder is interposed. The deliverer gives the customs modalities and names to the customs clearance forwarder. He generates the delivery note, conducts the transport to the customs clearance forwarder and subsequently to the customer. A goods receipt invoice is generated and directly sent to the customs clearance forwarder, since the deliverer must not come to know the customer price. The customs clearance forwarder possibly generates the necessary customs documents and carries out an import customs clearance.

As soon as the invoice is given from the deliverer to the invoice verification and worked out, the system takes over the invoice receipt amount in the customer invoice. With the invoicing of the third party order the delivery amount is automatically adopted in the customer invoice document.

The sale over a third party order to industrial consumers includes all important sales and distribution processes, in order to sell products by a deliverer to the customer. In particular this are the handling of customer orders with or without reference to quotations or outline agreements, the connection to the purchase as well as the invoicing. Further more the scenario includes different processes for the handling of customer reversals, subsequent payments like rebates or commissions.

During the different procedure types of the third party single process steps can be reduced to a minimum or totally cancelled.

The scenario is exclusively occupied with the operating procedure of products, which are sold by a deliverer. A direct reference between sales order and procurement does exist, the lead-time of the operating transaction is consequently in general relatively short.

Further on, the scenario is characterized by the standardization of the customer as an industrial consumer unlike the trading companies, end consumers.

In the scope of the third party processing, following R/3 components are integrated, as the following table shows:

R/3-component	Functions
Sales support	Customer acquisition
Sales	Customer inquiry,-quotation,- order procedure
Credit management	Credit limit check; assurance of open demands
Foreign Trade	Export control; Reports to authorities
Invoicing	Invoice-, credit/debit memo-, rebate procedure
Information system	Planning, Forecast, Statistic

Table 5.1: R/3 component applied in a third party business

5.1.2   Procedure of a third party business

Sales support

1. The customer order processing starts with the acquisition of customer order in scope of the customer contract processing (e.g. telephone calls, visits, letters)

2. The mailing action here offers the possibility of sending product information, marketing presents or models to selected customer.

Customer inquiries

3. The customer demand for a general product information (e.g. product description, availability) can be registered with the help of a quotation

4. Exact, compulsory product information (e.g. prices, delivery dates) is transmitted to the customer in form of a quotation. Out of a valid quotation accepted by the customer, an order can be directly generated.

Outline agreement

5. Long termed agreements for the acceptance of products in a fixed amount or a fixed value, are defined as contract. Out of contract recalls of the customer, concrete orders are established out of a contract.

Customer order

6. With a customer order short term agreements for the delivery of products to a customer are determined, in the course of which following central functions must be available for the company:

   - Calculation and pricing

   - Generating of an order requirement for the procurement of customer request materials

   - Dispatch termination

   - Credit limit check

   - Export control

   - Credit management

7. In the scope of the credit management different types of credit limit checks (e.g. statistical, dynamical) at different times of the order procedure (e.g. order, delivery, goods issue) can be made.

   In the risk management additional various function (e.g. document business, trade credit insurance) are available for securing the credit risk.

Foreign trade

In the foreign trade processing following functions are supported by the SAP R/3 system:

8. Export control on basis of permissions of order, deliverer or goods issue

9. Registration at institutions (e.g. INTRASTAT) on basis of customer invoicing by automatic registration procedures.

Invoicing

10. The invoicing generation takes place on grounds of the incoming invoice of the deliverer or on base of the dispatch notification of the deliverer. In the scope of reversal and volume-based rebate handling. Credit and debit

memos are generated. Invoicing plans support the user in the periodical or partial invoicing. The invoice processing results in the procedure of postings to the corresponding financial accounting –or cost accounting.

11.   With the help of an invoice list, lists of billings can be sent at a certain date to a payer.

Rebate settlement

12.   During the rebate settlement a credit memo is generated on base of all rebate relevant invoices, credit- and debit memos, by taking into account the agreements made. Rebate partial payments are possible and taken into consideration in the final settlement.

Information system

13.   With the help of the information system the system generates on demand plans, prognoses and analysis resulting of the business processes.

5.1.3   Result

Should returnable packaging be combined in a delivery, the scenario returnable packaging shows up.

For a possible required reversal procedure, the scenario returnable packaging can be considered.

5.1.4   Examples of third party deals

The customer places an order for material M to company XY. The company XY places this order to an external deliverer ED, who sends the material M directly to consumer C and passes the invoice to company XY. The customer on the other hand receives the invoice from company XY and also pays to XY.

Within the entry of an order in the SD module, which includes one or more third party positions, so called purchase requisitions are generated in the module MM, which there also become an order. The deliverer ED delivers the material M to consumer C. Referring the deliverer invoice, the invoicing in MM takes place, which on the other hand is the base for the invoice in SD. The generating of invoice in SD depends on invoice relevant indicators in the customizing. Here e.g. it can be controlled, whether the order should be immediately taken over in the invoice inventory stocks. Because of the difference between customer order quality and invoice quality the level of deliverance can be determined.

The type of material controls the third party processing. Different material types are related to the materials in accordance with its usage in the company. Every material can merely have one material type. The material type determines whether a material can be internally produced or only externally obtained. In general e.g. the trading good can only be externally obtained by a deliverer. if a material is only externally obtained, this can be fixed in the material master data. This material adequately is then decelerated as a so-called third-party position. The R/3 system then automatically determines the matching position type TAS.

The material type controls among others also the account determination and therefore determines which accounts are to be posted referring the accounting.

5.2   Returns and reversal procedure

5.2.1   Definition

A return is a return delivery of goods from the customer to the deliverer because of reset or scrapping on grounds of legal quality defects, false orders of the customer or false delivery of the deliverer with the demand for allowance of a refund amount or subsequent delivery free of charge. Because of the return the falsely ordered, falsely delivered or quality reversed good again passes into the deliverer's possession. Therefore the good can be examined in regards of quality defects and, concerning the further usage of the good, settled. With a legal reversal, the return initiates a subsequent delivery free of charge, referring the amount sent back and referring the invoiced value of the good a return credit memo. The good sent back becomes again the deliverer's property and is further used corresponding to the quality.

The reversal is a customer objection to a good or prices of a delivery. Even in the field service a reversal can be initiated. Grounds for a reversal are e.g. defect material, false deliveries, false orders or return deliveries before the deadline for returning goods in case of purchase on trail [Curran/Keller 99].

The scenario return and reversal procedure is represented as follows:

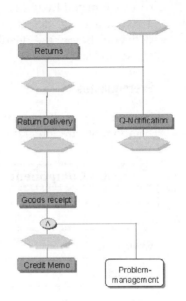

Illustration 5.2: Diagram return and reversal procedure

5.2.2   Usage/function

The reversal procedure starts with the customer reversal at the
customer consultant/field of service. Here first of all is to be
tested who is responsible. Then on grounds of the customer information it is tried to send a clarification of the reversal. Is it a
misordered, misdelivered or quality reversal good and should
this return into the deliverer's property, a return is initiated, if the
reversal is legal. This on the other hand results in a credit memo
in reference to the amount sent back and the invoiced value of
the (book credit memo/debit memo procedure) or in a subsequent delivery free of charge in reference to the amount sent
back. If the customer wants to keep the good and claims a price
reduction or a replacement free of charge, it must be decided
whether it can be allowed. Is this the case a credit memo or a
subsequent delivery free of charge on base of a SAP bonus
agreement can be initiated (also see bonus procedure).

5.2.3   Usage possibilities

This scenario describes the return and reversal procedure, which can be initiated by the customer or the field service. Purposes for a return are e.g. defect goods, misdiliveries, misorders or return deliveries before the deadline for returning goods in case of purchase on trial.

5.2.4   Prerequisites

In the scope of the return- and reversal procedure, following R/3 components are integratively used:

R/3 Component	Functions
Sales support	Reversal acceptance, Recall action
Reversal	Return, Credit memo demand
Dispatch	Return delivery, Goods receipt
Warehouse management	Storage of material
Quality management	Quality check for the return
Transport	Transport disposition,- handling
Invoice	Credit memo

Table 5.2: Used R/3 components with the return-and reversal procedure

5.2.5   Procedure

A reversal can be registered in the scope of a customer contact (e.g. telephone). This information serves then as base for the following reversal procedure.

If, for example, a recall action on grounds of product defects, determined in the scope of the quality assurance, is to be done, a mailing action is set for the notification of the concerned customer groups.

There are two possibilities for the procedure of reversals:

1. The customer sends back the good:
   Here a return is created in order to proceed a transaction if the good sent back has again reached the warehouse, a return delivery is created referring the already created return. The goods issue posted for the return delivery records the goods receipt to the own stock.

   After the goods receipt the return inspection takes place. Here, the return's admissibility is inspected and the good's usage decision determined (e.g. revision, scrapping).

   If the customer wants the paid amount back, a credit memo receipt referring the return is created. The SAP system then generates out of a granted credit memo requirement a credit memo for the customer, if required.

If the customer wants the goods replaced a subsequent delivery free of charge with reference to the return is created.

2. The customer doesn't send back the goods:
   If the customer wants the corresponding amount repaid a credit request with reference to the customer order is to be created. Out of an authorized credit memo request then a credit memo for the customer is generated.

If the customer wants the goods to be substituted a subsequent delivery free of charge in reference to the customer order is created.

5.3   Consignment procedure

5.3.1   Description

The sale from stock to the consignment taker includes all necessary processes of the sales in order to sell products from stock to the customer, who themselves lay the product first of all in their consignment warehouse. In detail these are the customer order processing with or without relation to quotations and outline agreement, the delivery and dispatch procedure, as well as the invoicing. Further more, the scenario includes processes for the procedure of customer reversals, subsequent allowances like volume-based rebates or provisions as well as returnable packaging or empties (packaging, or the like).

With the different types of sales from stock procedure single process steps can be reduced to a minimum or totally cancelled. As e.g. the variant cash sales, which in general proceeds without the dispatch procedure, since the customer takes the good directly along.

The scenario deals extraordinarily with the operational/business procedure of products, which are sold from stock. Consequently serial products are concerned, which have been produced or purchased on grounds of sales and quantity planning, which can be consumption or prognoses orientated repeatedly ordered or directly delivered from stock to the customer.

There is no direct connection between sales order and procurement; the lead-time of the sales transaction is therefore in general comparatively short. Immaterial goods as services are also dealt with at another place as well as all variants of a product of customer orientated products or final assembly of products. Further more, the scenario is characterized by the typology of the customer as an industrial consumer in difference to trading companies and (final) consumer; with the exception of the typical procedure in the supplier, which is also described in an own scenario.

The scenario consignment procedure is graphically described as follows:

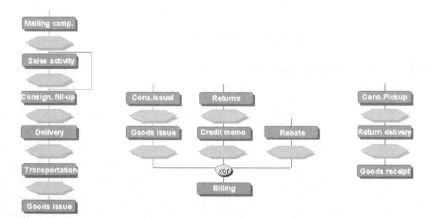

Illustration 5.4: Diagram consignment procedure in general

5.3.2   **Definition consignment**

Consignment means, that a deliverer makes a material available for a consumer, which is stored by the consumer. The deliverer is as long proprietor of the material, as the consumer produces from the consignment store. Therefore a liability towards the deliverers comes into existence. The invoice is paid after arranged periods e.g. monthly. The contract partner can agree upon the take over of the remaining consignment stock on part of the customer into the own stock.

Advantages of consignment:

- The consignment stock is managed among the same material number as the own stock. With that, the consignment stock can be integrated in the stock available of a material

- Consignment stocks of the same material of different deliverers are managed independently from each other and with the particular price of the deliverer

- Consignment stocks are not valued. A taking out is valued with the price of the respective deliverer.

- A consignment material can be registered in three stock types

  - Freely usable stock

  - Quality test facility

  - Blocked stock

Transfer postings are possible between all stock types. Withdrawals are only allowed from freely available stock.

Before a deliverer orders a material or a goods receipt in the consignment stock is posted, the price of the deliverer must be maintained. The price is necessary for the calculation and billing of the material. If several deliverers order consignment material, a consignment stock is administrated for every deliverer, since the particular amount of the material can have different prices.

Goods receipts in the consignment can either be referred to an order or directly ensue from the deliverer without order.

After the goods receipt of a consignment material the consignment stock of the material increases. The consignment stock is available for the disposition.

Thus the valued stock of the material doesn't increase, since the consignment stock remains in the deliverer's property.

The billing of consignment goods ensues without invoice receipt, since the deliverer can't directly observe the goods withdrawal he can administrate the consignment stock merely through his deliveries and payments.

5.3.3   Prerequisites

The quality of a material or product is to be inspected, before it leaves the factory site of the producer or deliverer. Most of the entire department responsible for the quality assurance precedes a quality check, in order to ensure, that the goods are impeccable, before sent to the customer.

A consignment good is a good stored by the customer but remains in the property of the delivering company. The customer is only then obliged to bill the good with the deliverer, when he withdraws it from the consignment stock. Otherwise he can return unnecessary consignment goods. Since the consignment stocks are part of the deliverer stock valued, of these must be lead into the single of the deliverer.

Nevertheless, the consignment stocks must be separately kept from the other stock, in order to gain an overview of the customer's store. Furthermore they are to be separately kept for different customers.

5.3.4   Scenarios

In the scope of the scenario sale from stock to consignment taker the following R/3 components are to be integrated the best.

R/3 component	Functions
Sales support	Customer acquisition
Sales	Customer inquiry,-quotation,- order procedure
Credit management	Credit limit check; Assurance of open demands
Dispatch	Delivery procedure, goods issue
Warehouse management	Goods transfer/release to/from stock
Quality management	Quality check for the delivery and return
Transport	Transport disposition and termination
Foreign trade	Export control; report to authorities
Invoice	Invoice-, credit/debit memo-, rebate procedure
Information system	Planning, forecast, Statistics

Table 5.3: R/3 components used with a sale from stock to consignment

5.3.5   Procedure

Sales Support

1. The customer order processing starts with the acquisition of customer orders in the scope of customer contact processing (e.g. telephone, visit, letter)

2. The mailing-action offers possibility to send product information, promotion gifts or models to a selected customer type.

Customer order

3. Three different order types are relevant for the consignment procedure:

   - The filling of the customer consignment warehouse ensures with a consignment fill-up order

   - If the customer withdraws goods out of the consignment stock for usage or sale, this transaction can be fixed by creating a consignment withdrawal.

   - Returns of goods off the consignment stock are made by a consignment pickup

   Following standard functions of the order processing are available for the consignment procedure:

   - Calculation and pricing

   - Availability check

   - Dispatch termination

   - Passing on of requirements to program planning

   - Credit limit check

   - Export control

   - Credit management

Credit management

4. In the scope of the credit management several types of the credit limit checks (e.g. statistical, dynamical) can be gone through to several terminations of the order processing (e.g. order, delivery, goods issue).

   In the risk management additional different procedures (e.g. document business, payment cards, goods credit assurance) are available for the assurance of the credit risk.

Dispatch

5. The dispatch procedure is initiated by the establishment of deliveries from dispatch due data orders. Herewith the availability situation and the scheduled shipping date validity are again checked.

6. Subsequently, the goods withdrawal and the provisions are processed with in the scope of the picking. The picking can be made by the support of the warehouse management system.

7. The packing of delivery positions ensues by allocation of dispatch elements, which can be a combination of materials and packaging material or again dispatch elements.

8. The goods issue marks the end of the dispatch processing. It differently proceeds in dependence of the based order:

   • Consignment fill-up-order:
     The adequate amount is directly debited from the regular stock of the deliverer site and is added to the customer special stock. The total valued stock of the deliverer site remains unchanged

   • Consignment withdrawal:
     With the posting of the goods issue the corresponding amount is drawn off the customer special stock as well as the own stock valued.

   • Consignment pickup:
     With the posting of the goods issue the corresponding amount is directly debited from the customer special stock and posted to the normal stock of the site, to which the good is returned. The total deliverer stock value here stays unchanged, since the good sent back has been part of the customer's stock, too.

Warehouse management

9. The combination with the warehouse management system (WM) ensues in accordance with the picking. Herewith it is differentiated between the picking in fixed placed warehouses and chaotically organized warehouses (usage of transport orders).

10. If creating a delivery note for a product, the R/3 system automatically opens an inspection lot and makes the necessary work documents available for the quality assurance. There's also the possibility, of manually creating the inspection lot.

    If the product to be delivered is posted in batches and a delivery position can't be satisfied within a single batch, the R/3 system can divide the delivery position into several batches and generate a partial log for every batch.

If it's determined, that the quality assurance can ensue after the delivery, the goods issue can be independently posted from the quality assurance.

11. Possible faults can be documented by failure data. If the connected adjustment measures should be managed in the R/3 system, quality information is generated in the scope of the problem management and deal with.

12. If the feature results are registered in partial lots, the usage is determined for the partial lots and the total inspection lot. Inspection activities, which occurred in the scope of the result registration or with the usage decisions made, can at this moment be collected on a QM order.
The system automatically creates a quality certificate if required.

13. If the quality inspection hasn't shown a result, the goods issue can be posted after the usage decision made.

Transport

14. In the transport management the transport due deliveries can be summarized in transports. Essential management activities are the determination of dispatch dates, the determination of transport tools and the route assignment.

15. The transport handling enables a follow up of the transport activities to be made after the management.

Foreign trade

In the foreign trade procedure the user is supported by following functions:

16. Export control on base of permits in order, delivery or goods issue.

17. Information to authorities (e.g. INTRASTAT) on base of customer invoice by an automatic reporting procedure.

Invoicing

18. The billing follows up on grounds of deliveries (products) or orders (services). The invoice determination results in the execution of postings on the corresponding financial accounting –and cost accounting accounts. In accordance to the consignment procedure only the consignment withdrawal is invoice relevant, since here a property transit takes place.

19. With the help of a billing list a list of invoices can be sent to a special date to a payer.

Credit memo procedure

20. With the rebate settlement a credit memo is created on base of all rebate relevant invoices, credit and debit memos with consideration of decisions made in the rebate agreement. Rebate partial payments are possible and taken into consideration in the final settlement.

Information System

With help of the information system planning, prognoses and analyses are created on base of master data and the transaction data resulting of business processes.

5.4.   Processing of the customer consignment fill-up order

5.4.1   Definition

Customer consignment fill-up order means the dispatch of goods to a consignment store of the customer

5.4.2   Usage/task

Does no customer special stock exist in the stock, this is automatically created by the posting of the goods issue. The corresponding amount is booked out of the regular stock in the customer special stock. The consignment stock stays in the property of the company.

5.4.3   Description

The consignment fill up order is registered in the system with an own order type KB. It is to be arranged organizationally, that consignment fill-up orders can be created for those customers, who do have a consignment warehouse contract. The consignment fill-up order can be differentiated from the standard order as follows: Prices are not registered for the product to be delivered, since the prices are only applied for invoicing, when the goods are withdrawn. So there is no value in the order. With that no credit limit check can ensue. But this is not necessary, since consignment warehouse contracts are merely made with solvent customers. The order registration is followed by delivery and after the picking the goods issue is posted. There, no invoicing takes place, since the stock is up to now in the property of the deliverer. Here the problem is posed how information about the existing freight cost gets to the invoice verification for the inspection of the incoming freight invoices.

5.5   Procedure of customer consignment withdrawal

5.5.1   Definition

Customer consignment withdrawal deals with the property transit of goods to the customer and the invoicing of goods.

5.5.2   Usage

After the customer's withdrawal of the consignment stock for use or sale, a withdrawal confirmation is made at the end of the month. The corresponding amount is drawn off the customer special stock as well as off the valued total stock by booking. The good is now billed to the customer.

5.5.3   Description

The consignment withdrawal order is registered with a special order type (KE) in the system. With the registration reference can be made to the consignment fill-up.

The order registration does hardly differ from the one of the standard order. Thus, goods are given from the customer stock. Hereby it is inspected against the consignment stock of the customer. The current pieces are taken as a basis.

On behalf of the transport costs it is to be considered, that these have already come up in the consignment fill-up (for the total amount of the fill-up), but are only now passed to the customer's account (caution: freight costs must not be referred to the withdrawal registered amount, since an authorized surcharge for quantities below minimum can be billed; it must rather be distributed to all withdrawals).

A credit limit check is not made in this place, since the good is already delivered and if necessary used.

Now the delivery follows. Here a batch is registered. Since the customer has no batches in his consignment warehouse, he doesn't give those with his withdrawal report. Here, the procedure 'first in-first out' can be applied, that means the registrant chooses a batch and views it with the order.

After the goods issue counting the invoicing takes place. In the invoice also the information of the package of material must be included (for package disposal).

5.6 Procedure of customer consignment pickup

5.6.1 Definition

Customer consignment pickup means the redelivery of goods from the customer consignment warehouse. Up to now the good is invoiced to the customer.

5.6.2 Usage

Starting point for a pickup from the customer consignation warehouse could be:

- The customer doesn't need the good any more (it is merely happening e.g. in case of a bankruptcy of the customer)
- The good is delivered to the wrong consignment warehouse (misdilivery – very seldom)
- Goods, which don't meet the customer expectations, are taken back by deliverers in considered cases

5.6.3 Description

For all cases mentioned above counts: If the pickup is authorized by the marketing, a consignment pickup order is created (order type K1), with reference to the consignment fill-up.

The good is either sent back from the customer to the deliverer or picked up by the deliverer from the customer. In case of reversed goods it can happen, that the customer takes over the disposal by his own.

In the pick-up order all data necessary about the customer, the consignment good and the site, to which the good is returned, must be registered. Then a return delivery is created to the order. The goods issue posting concludes the transaction and initiates a posting from the customer special stock to the site stock.

The good is not posted in the stock, if the customer disposes the good by his own. The special stock must be correspondingly booked.

The product does not change due to the return, the type stays the same, only the quality changes. After the quality inspection made, there a following possibilities:

- Material corresponds to the specifications, that means, posting in the freely usable stock

- Material does not (any longer) correspond to the specification but is still salable

Amount differences can occur with returns. These amounts are to be registered as goods receipt by MM.

Is a less amount sent back, because of a taking of samples for inspection of the reversal, the differences must also be corrected in the stock. The consignment stock, which should exist, is completely posted out. Does the customer dispose the good by his own, the stock must be booked out of his site.

With transfer deliveries or misdeliveries, a new order is generally created after the goods return, for the dispatch to the right consignment stock or to a third customer.

The consignment pick-up is not invoice relevant, since the good is still in the property of the company. Here the question is to be posed of how the transport costs caused by the business transaction can be transmitted to the CO.

5.7   Procedure of customer consignment returns

5.7.1   Definition

Customer consignment return means the return delivery of goods, which were part of the customer consignment stock, were taken out from it and therefore have already been invoiced. Since the good returns in the property of the original enterprise, the transaction is invoice relevant. The customer receives a credit memo for the goods sent back.

5.7.2   Purpose/tasks

Since the goods go back into the property of the original delivering company, the transaction is invoice relevant. The customer gets a credit memo.

5.7.3   **Description**

Since the good is already invoiced, this is an ordinary return. The procedure is therefore identical with a common return procedure. But there is a specific order type. The consignment withdrawal can be referred to. In the scope of the creation of a consignment return an invoice blocking is automatically set. In order to invoice this return, first of all the invoice blocking has to be removed, surely after it has been proved if a credit memo is legal.

With the return it has to be considered, that the credit memo has to include information of the package (for the package disposal). Further more the same text as in the invoice sales must be used.

5.8.   Returnable packaging and empties procedure

5.8.1   **Usage possibilities**

The returnable packaging are articles, which lie in the customer stock, but are still property of the delivering company. The customer is only then obliged to invoice the returnable packaging with the deliverer if he has not sent it back to a specific date. So it is possible, to proceed the invoicing respectively the return of pallets or empties with this function. Even the resale of returnable packaging to third parties can be regulated in this way. Since the returnable packaging are part of the deliverer's evaluated stock, these must be lead in his system [Curran/Keller 99/169].

The returnable packaging stock must be separately kept, for the deliverer can keep an overview of the customer's stock and of different customers.

The scenario returnable packaging and empties processing is graphically depicted as follows:

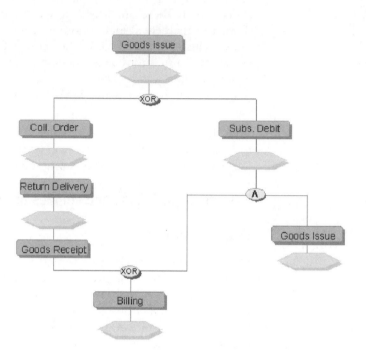

Illustration 5.7: Diagram scenario returnable packaging and empties procedure

5.8.2   Prerequisites

The scenario returnable packaging/empties procedure only deals
with the returnable packaging pick-up and the charges for packaging not returned. The returnable packaging fill-up (order)
takes place in the scope of the common customer order procedure and is only mentioned here for the sake of completeness.
The scenario can therefore only be applied in combination with
other sales/distribution scenario.

With the use of returnable packaging/empties following R/3 components are integrated applied:

R/3 component	Functions
Sales	Customer order processing
Dispatch	Delivery procedure, goods issue/receipt
Transport	Transport disposition, - scheduling
Invoice	Debit memo processing

Table 5.4: R/3 components used within a returnable packaging and empties procedure

5.8.3   Procedure

The returnable packaging/empties processing consists of the three following separate business processes:

- Returnable packaging fill-up
- Pick-up of returnable packaging
- Charges of packaging not returned

Fill-up of returnable packaging

Customer order

1.   A returnable packaging is separately positioned in the customer order, whereas following central functions are available:

    - Availability check
    - Dispatch determination
    - Demand transmission to the program planning

Dispatch

2.   Dispatch handling is initiated by the creation deliveries of orders due for dispatch. Hereby the availability situation and the dispatch validity are newly checked.

3. Ensuing, the goods withdrawal and the dispatch determination validity are processed in the scope of the picking.

4. The completion of the dispatch handling is marked by the goods issue. If yet there doesn't exist a special stock for the customer in the stock of the deliverer (or special stock lead), then this is automatically generated with the posting of the goods issue. The corresponding amount is posted off the regular stock of the deliverer plant and added to the customer special stock. The total stock evaluated of this plant stays hereby completely unchanged.

Transport

5. Deliveries due to transport can be summarized to transports in the transport disposition. Essential disposition activities are the determination of dispatch deadlines, the calculation of transport means and the route allocation.

6. The transport handling enables a follow-up of the transport activities to be undertaken after the disposition/planning

The transaction is not invoice relevant, since the customer special stock remains property of the delivering company.

Returnable packaging pick-up

1. The return of returnable packaging is handled by a pick-up of returnable packaging. Therefore a corresponding order (returnable packaging pick-up) with the returnable packaging position is generated. Due to the application of the availability check a survey of the returnable packaging can be achieved.

2. The dispatch handling registers the returnable packaging picked-up or delivered as return delivery.

3. With the posting of the goods issue the corresponding amount is posted off the special customer stock and posted to the general stock of the plant, to which the good is returned. The deliverer's total stock evaluated stays unchanged, since the good returned was always part of the deliverer's own stock.

The transaction is not invoice relevant, since the customer special stock is in the property of the deliverer.

Charges for packaging not returned

1. The customer is charged with the returnable packaging if he wants to keep it or has damaged it. Here a corresponding order (charges of packaging not returned) is created with the returnable packaging positions.

2. The dispatch handling generates a delivery for being able to post the goods issue.

3. With the posting of the goods issue, the correlating amount is withdrawn from the customer special stock and from the own evaluated stock.

4. The deliverer generates an invoice for the delivery since the returnable packaging has transferred into the customers' property

5.9   Presales procedure (sales to industrial customer)

5.9.1   Usage possibilities

This scenario describes the definition of master data being the prerequisites for the sales scenario handling for industrial customers.

5.9.2   Prerequisites

The sales specific views of the master data processes from the material to the customer are located in the scenarios of the business process fields product development, marketing and business partners. They are obligatory for the processing of this scenario.

5.9.3   Procedure

Material maintenance

1. Material combinations often ordered by a customer, are the best represented with the help of collections. These collections can be applied as position proposals in the order handling.

2. The material finding serves for the automatic material replacement or the display of a selection list of materials in the order handling. An example for this is the substitution of normal goods by campaign goods.

3. If materials authorized for the customer should be limited, they are placed in the material list exclusion.

Conditions

4. Pricing elements important for the pricing (e.g. pricelist, material prices, customer individual price) are determined in the scope of the condition processing. Further more, the corresponding surcharges and advanced payments are defined in dependence on different criteria (e.g. customer, customer hierarchy, material). Conditions are also applied by the application of taxes.

Agreements

5. So called 'customer-material-information' is applied for the definition of customer specific agreements for certain materials (e.g. delivery plant).

6. Sales dependent discounts for customers are determined as rebate agreements. The conditions necessary for the rebate evaluation can be defined in dependence on several criteria (e.g. customer, customer hierarchy, customer/material).

Batch administration

7. In the scope of the batch administration, the criteria to be applied (e.g. customer, material) for the establishment of the batch, are generated with the search strategy in a business process dependent way (e.g. delivery).

Availability check

8. If a periodical specific division of limited products in dependence of special criteria (e.g. customer, region) should be realized, then the quota processing should be used the best. This division is considered in the availability check.

Foreign trade

9. Here, the master data necessary for the automatic export control and the information to authorities is generated.

10. The calculation necessary for the preference procedure is made and the financial document master record for the L/C procedure is defined.

Message finding

11.   The term message finding includes both the electronic and the printed documents and trading papers as e.g. order acknowledgement and delivery note, but also internal mails. The control of the message exchange (e.g. time and type of the message transfer) can be adjusted company specifically.

Quality management

12.   If the application of certain production –or operation procedures, the processing of defined tests on basis of laws, norms, customer specifications or other agreements and if given proper test results should be stated, the basic data are generated for the apply of quality certification. This is proceeded as follows.

    a.   A formula is generated in the scope of the quality certification planning. It controls the layout of the pages and the format of the data in the certification

    b.   Within the quality certification planning a certification model is created. A certification model controls the selection of inspection lots and partial lots and the selection of the features. The certification models must be released before usage.

    c.   Subsequently, the certification model is allocated to another object, that means e.g. to a combination of material and customer.

    d.   Finally possible recipients of a certificate are to be defined.

If the R/3 system should react automatically to the realization of delivery positions, the sales relevant CONTROL DATA IN QM must be maintained. Herewith the quality check of a delivery can be influenced, among others based upon following parameters:

- Material
- Customer
- Delivery

5.10   Business partner handling on the debits side

5.10.1   Usage possibilities

A company has, in accordance to its business field, to deal with different natural and legal persons: A customer orders goods from the company. For the proper delivery to the customer, a forwarding agent is, if necessary, charged with the delivery. A sales employee processes the business transactions in the company. All roles possibly occupied by a natural or legal person are represented in the SAP system as business partners. The subsequent description refers exclusively to the business partners on the debits side.

5.10.2   Procedure

Customer master

1. The customer master includes data of the business partners, with whom the company maintains business relationship with regard to the delivery of products and/or services. A separate customer master record is created for every customer (except CPD-customers). The customer data is divided in three categories, which are administrated in one common master data record:

    *   General data

    *   Sales specific data

    *   Company code specific data

    The company structure of the customer can be complex and different company departments can be responsible for different functions. This fact is represented by different roles, which can be hold by a customer:

    *   Ordering party

    *   Invoice recipient

    *   Payer

    *   Goods recipient

2. Personnel data of representatives or sales managers are maintained in personnel master data records in the HR. Herewith e.g. a direct commission payment is possible.

3. With the help of customer hierarchies flexible customer structures, (e.g. purchasing administration) can be represented and be used in the scope of the pricing and for statistical surveys.

Classification

The R/3 category system is a generic tool for the classification of data objects and includes classification and search criteria for materials, customers and deliverers. Data objects can be allocated to different categories with various features.

5.11   Delivery of models and promotion means

5.11.1   Usage possibilities

Model and promotion means are in general delivered or sent free of charge to the customer.

The self-production is ahead of the delivery of models and the procurement of promotion means, which are normally trading goods. The costs can be planned in the CO, in the financial statement or in the internal orders and then be evaluated with the actual values, which are given over in the invoicing.

Besides of the delivery by a field service employee, the articles can be sent also by delivery and subsequent transport or as addition to a marketing action.

Model and promotion means can, if possible also be invoiced with a price.

5.11.2   Prerequisites

In the scope of the customer order handling following R/3 components are integrated applied in the delivery of models and promotion means:

R/3 component	Functions
Sales support	Customer contact, mailing actions
Sales	Order for a delivery free of charge

R/3 component	Functions
Dispatch	Delivery processing, goods issue
Warehouse management	Transfer/release of materials
Transport	Transport disposition and termination
Foreign trade	Export control, reports to authorities
Invoicing	Invoice procedure; cost transfer to the result invoice
Information system	Planning, forecast, statistics

Table 5.5: R/3 components used in the delivery of models and promotion means

5.11.3   Procedure

Sales support

1. The customer order handling starts with the acquisition of customer orders in the scope of a customer contact processing (e.g. telephone call, visit, letter).

2. The mailing action offers the possibility to send product information, promotion gifts or models to a selected customer group.

Customer order

3. With a customer order the models and promotion means are registered as positions free of charge. They can also be registered mixed with other products liable to pay costs. Herewith following central functions are available.

   - Calculation and pricing
   - Availability check
   - Demand transfer to the program planning
   - Dispatch determination
   - Export control

Dispatch

4. The dispatch procedure is initiated by the criteria of deliveries of orders due to dispatch. Herewith the availability situation and the dispatch determination validity are again checked.

5. Subsequently the goods withdrawal and the provision in the picking area are made in the scope of the picking.

6. The packaging of delivery positions ensues by distributions of dispatch elements, which can be a combination of materials, packing material or again dispatch elements.

7. The termination of the dispatch procedure is marked by the goods issue, which leads to a reduction of the stock and actualization of the evaluation of the balance sheet accounts in the financial accounting.

Warehouse management

8. The combination with the warehouse management system (WM) takes place in the scope of the picking. Hereby it is to be distinguished between the picking in warehouse with fixed bin locations and chaotically organized warehouses (usage of transport orders).

Transport

9. In the transport disposition the deliveries due to transport can be collected in transports. Essential disposition activities are the determination of dispatch dates, the determination of transport means and the route allocation.

10. The transport handling enables a monitoring of transport activities to be made after the disposition.

Foreign trade

Following functions are supported in the foreign trade procedure:

11. Export control on base of permits in order, delivery or goods issue.

12. Reports to authorities (e.g. INTRASTAT) on base of customer invoicing by an automatic reporting procedure.

Invoicing

13.   The invoice creation is made on grounds of deliveries. It results in the procedure of postings to the corresponding financial accounting and cost accounting accounts, whereby here in general, only the costs are posted, not the revenues.

Information system

14.   With the help of the information system planning, forecast and analyses based on master and movement data referring to the business processes, are created.

5.12   Sale from stock by scheduling agreement

5.12.1   Usage possibility

The sale from stock via scheduling agreements deals exclusively with the operational procedure of products sold from stock. Consequently, these are serial products produced or picked on grounds of a sales and turnover planning, are reordered forecast orientated and can directly be delivered to the customer (off plant). There is no direct relation between sales order and picking, the lead-time of operational transaction is therefore relatively short. Immaterial goods as services are as well dealt with at another space as all variants of a customer related production or final assembly of products.

The scenario sale from stock via scheduling agreement is graphically described as follows:

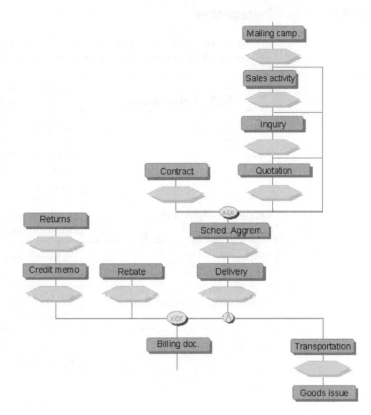

Illustration 5.8: Diagram sales from stock via scheduling agreement

The quality of a material or product is to be checked before leaving the site of the producer or deliverer. Mostly, the deliverer's department responsible for the quality assurance makes the quality check, in order to assure that the goods are in unobjectionable condition before sent to the customer.

With a delivery schedule an outline agreement over the product to be delivered on a specific date, is made. Since the delivery schedule already includes concrete delivery dates (distributions), the creation of the delivery directly ensues without further order registration.

5.12.2   Prerequisites

In the scope of the scenario sale from stock via scheduling agreement following R/3 components are used integrated:

R/3 components	Functions
Sales support	Customer acquisition
Sales	Customer inquiry-, customer quotation-, delivery schedule procedure
Credit management	Credit limit check, assurance of open demands
Dispatch	Delivery processing, goods issue
Warehouse management	Transfer/release of materials from/to stock
Quality management	Quality check referring delivery and return
Transport	Transport disposition and termination
Foreign Trade	Export control, reports to authorities
Invoicing	Invoice- credit/debit memo-, rebate procedure
Information system	Planning, forecast, statistics

Table 5.6: R/3 components used within the sale from stock via scheduling agreement

5.12.3   Procedure

Sales support

1. The customer order handling starts with the acquisition of customer orders in the scope of the customer contract handling (e.g. telephone, visit, letter).

2. The mailing campaign offers the possibility to send product information, promotion gifts or models to a group of selected customers

Customer inquiry/quotation handling

3. The inquiry of a customer about general product information (e.g. product description, availability) can be registered with the help of an inquiry.

4. Exact and liable product information (e.g. prices, delivery dates) is transferred to the customer in form of a quotation. An order can directly be generated out of a quotation, which is valid and accepted by the customer.

Outline agreement

5. Long term agreements for the acceptance of products in a fixed amount or a fixed value are defined as contracts. The agreements defined in a contract can be used as basis for the creation of a scheduling agreement.

6. The scheduling agreement offers the possibility, to determine already concrete delivery amounts- and dates for a special validity date. Every scheduling agreement distribution has the character of a customer order and is separately delivered by the due date of dispatch. Following central functions are available in the scheduling agreement:

- Calculation and pricing
- Availability check
- Dispatch determination
- Demand transfer to the program planning
- Credit limit check
- Export control
- Credit management/risk management

Credit management/risk management

7. With the credit management various types of the credit limit check (e.g. static, dynamical) can be performed at different times of the order processing (e.g. order, delivery, goods issue).

In the risk management various procedures (e.g. document business, payment cards, trade credit insurance) are available for the assurance of the credit risk.

Dispatch

8. The dispatch procedure is initiated by the creation of deliveries of orders due to dispatch. Herewith the availability situation and the validity of the dispatch deadline are checked anew.

9. Subsequently the goods withdrawal and the provision in the field of picking are proceeded in the scope of the picking.

10. The packing of delivery positions ensues by distribution of dispatch elements, which can consist of a combination of materials and dispatch materials or on the other hand of dispatch elements.

11. The termination of the dispatch procedure is marked by the goods issue, leads to a stock reduction and an actualization of the balance sheet accountancy evaluation in the financial accounting.

Warehouse management

12. The combination with the warehouse management system takes place in the scope of the picking. Herewith it is distinguished between the picking in warehouses with fixed bin locations and those, which are chaotically organized (usage of transport orders).

Quality assurance

13. If a delivery note is generated for a product, the R/3 system *automatically* opens an inspection lot and provides the business documents necessary for the quality assurance. Nevertheless the inspection lot can also be *manually* generated.

If the product to be delivered is performed in batches and a delivery position is not to be satisfied in a single batch, the R/3 system can divide delivery positions in several batches and generate a partial lot for every batch.

If it is determined that the quality assurance can follow after the delivery, the goods issue can be posted independently from the quality assurance.

14. Feature results are registered for the inspection lot respectively partial lot.

15. Possible failures can be documented by failure data if the adjustment measures combined shall be managed in the R/3 system, a quality information is to be made in the scope of the problem management and its performance.

16. If the feature results to the partial lots are registered, the usage decision for the partial lots and the total inspection lots is to be made. Otherwise a usage decision is made for the inspection lot.

Costs, incurred in the scope of the result registration or the usage decision, can be at those times confirmed to the production planning. The system automatically generates a quality certification if required.

17. If the quality check shows no result, the goods issue can be posted after the usage decision (has been made).

Transport

18. The deliveries due to dispatch can be collected in the transport disposition to transports. Essential disposition criteria are the determination of dispatch deadline, the calculation of transport means and the route allocation.

19. The transport handling enables the follow up of the transport activities to be made after the disposition.

Foreign trade

Following functions do support the customer in the foreign trade procedure:

20. Export control on base of permits in order, delivery or goods issue.

21. Reports to authorities (e.g. INTRASTAT) on base of customer invoicing by an automatic reporting procedure.

Invoicing

22. The invoice creation is made on grounds of deliveries (products) or orders (services). Invoicing plans support the periodical or partial invoicing. The invoice creation results in the procedure of postings to corresponding financial accounting and cost accounting accounts.

23. With the help of an invoice list, lists of invoices on a deadline to a payer can be made.

Rebate procedure

24. With the rebate billing a credit memo is generated in consideration of all rebate relevant invoices, credit and debit memos and the arrangements made in the rebate agreement. Rebate partial payments are possible and considered in the final settlement.

Information system

25. With the help of the information system planning, forecasts and analysis are generated on base of master data and resulting movement data of business processes.

5.13   Sale from stock to industrial consumers

5.13.1   Usage possibilities

The sales form stock to industrial consumers deals exclusively with the sales procedure of products sold from stock. Thus it refers to serial products, produced or procured on grounds of sales and operations planning, which can be reordered demand and forecast orientated, directly be delivered from stock to the customer. There is no direct reference between sales order and procurement, which shortens in general the lead-time of a sales transaction. Immaterial goods as services are also dealt with elsewhere as all variants of a customer related production or final assembly of product.

The quality of a material or product is to be tested before leaving the site of the producer of deliverer. Most of the time the quality check is made by the deliverer's department responsible for the quality assurance, in order to assure that the goods are in perfect condition before sent to the customer.

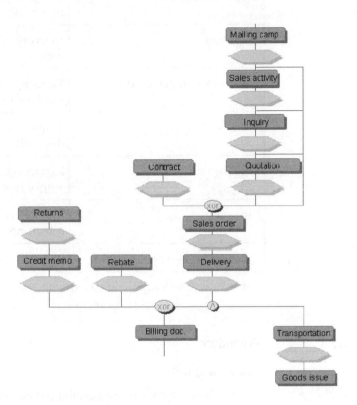

Illustration 5.9: Diagram sales from stock to industrial consumer

5.13.2   **Prerequisites**

Following R/3 components are integrative used with in the scenario sale from stock to industrial consummates.

R/3 components	Functions
Sales support	Customer acquisition
Sales	Customer inquiry, customer quotation-, order procedure

R/3 components	Functions
Credit management	Credit limit check, insurance of outstanding demands
Shipment	Delivery processing, goods issue
Warehouse management	Transfer and release to/from stock of materials
Transport	Transport disposition and dispatch
Foreign trade	Export control, reports to authorities
Invoicing	Invoice, credit-/debit memo-, rebate processing
Information system	Planning, forecast, statistics

Table 5.7: R/3 Components used with the sale from stock to industrial consumers

5.13.3   Procedure

Sales support

1. The customer order handling starts with the acquisition of customer orders in the scope of the customer contract handling (e.g. telephone, visit, letter).

2. The mailing campaign offers the possibility to send product information, promotion gifts or models to a group of selected customers.

Customer inquiry/quotation handling

3. The inquiry of a customer about general product information (e.g. product description, availability) can be registered with the help of an inquiry.

4. Exact and liable product information (e.g. prices, delivery dates) are transferred to the customer in form of a quotation. An order can directly be generated out of a quotation which is valid and accepted by the customer.

Outline agreement

5. Long term agreements for the acceptance of products of fixed amounts or a fixed value are defined as contract. With customer contract releases concrete orders a generated off a contract.

Customer order

6. The scheduling agreement offers the possibility, to determine already concrete delivery amounts- and dates for a special validity date. Every scheduling agreement distribution has the character of a customer order and is separately delivered by the due date of dispatch. Following central functions are available in the scheduling agreement:

    *   Calculation and pricing
    *   Availability check
    *   Dispatch determination
    *   Demand transfer to the program planning
    *   Credit limit check
    *   Export control

Credit management

7. With the credit management various types of the credit limit check (e.g. static, dynamical) can be performed at different dates of the order processing (e.g. order, delivery, goods issue).
    In the risk management various procedures (e.g. document business, payment cards, trade credit insurance) are available for the assurance of the credit risk.

Dispatch

8. The dispatch procedure is initiated by the creation of deliveries of orders due to dispatch. Herewith the availability situation and the validity of the dispatch deadline are checked anew.

9. Subsequently the goods withdrawal and the provision in the field of picking are proceeded in the scope of the picking.

10.   The packing of delivery positions ensues by distribution of dispatch elements, which can consist of a combination of materials and dispatch material or on the other hand of dispatch elements.

11.   The termination of the dispatch procedure is marked by the goods issue, leads to a stock reduction and an actualization of the balance sheet account evaluation in the financial accounting.

Warehouse management

12.   The combination with the warehouse management system takes place in the scope of the picking. Herewith it is distinguished between the picking in warehouses with fixed bin locations and those, which are chaotically organized (usage of transport orders).

Quality assurance

13.   If a delivery note is generated for a product, the R/3 system automatically opens an inspection lot and provides the business documents necessary for the quality assurance. Nevertheless the inspection lot can also be manually generated.

If the product to be delivered is performed in batches and a delivery position is not to be satisfied in a single batch, the R/3 system can divide the delivery position in several batches and generate a partial lot for every batch.

If it is determined that the quality assurance can follow after the delivery, the goods issue can be posted independently from the quality assurance.

14.   Feature results are registered for the inspection lot respectively partial lot.

15.   Possible failures can be documented by failure data. If the adjustment measures combined shall be managed in the R/3 system, a quality information is to be made in the scope of the problem management and its performance.

16.   If the feature results to the partial lots are registered the usage decision for the partial lots and the total inspection lots is to be made. Otherwise a usage decision is made for the inspection lot.

The system automatically generates a quality certification if required.

17. If the quality check shows no result, the goods issue can be posted after the usage decision (has been made).

Transport

18. The deliveries due to dispatch can be collected in the transport disposition to transports. Essential disposition criteria are the determination of dispatch deadline, the calculation of transport means and the route allocation.

19. The transport handling enables the follow up of the transport activities to be made after the disposition.

Foreign trade

Following functions do support the customer in foreign trade procedure:

20. Export control on base of permits in order, delivery or goods issue.

21. Reports to authorities (e.g. INTRASTAT) on base of customer invoicing by an automatic reporting procedure.

Invoicing

22. The invoice creation is made on grounds of deliveries (products) or orders (services). Invoicing plans support the periodical or partial invoicing. The invoice creation results in the procedure of postings to corresponding financial accounting and cost accounting accounts.

Rebate procedure

23. With the rebate billing a credit memo is generated in consideration of all rebate relevant invoices, credit and debit memos and the arrangements made in the rebate agreement. Rebate partial payments are possible and considered in the final settlement.

Information system

With the help of the information system planning, forecasts and analyses are generated on base of master data and resulting movement data of business processes.

5.14   Sale from stock to internal consumers

5.14.1   Usage possibilities

The sale from stock to internal consumers describes the goods movement between the production and distribution warehouses.

The distribution of the sales products to the regional or central warehouses, from which the customers are supported (with), results in the automatic creation of an order proposal in the preliminary scenarios planning and procurement, which then is transferred in the replenishment delivery of the site warehouses or the production.

Another usage possibility exists, if preliminary products are produced at a site and then have to be transferred to another site for further processing.

An internal charging has to be generated in case of different company codes of the delivery –and the receiving plant.

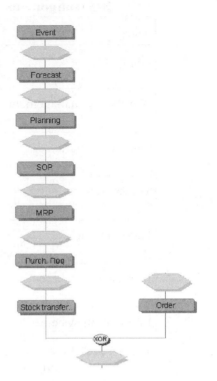

Illustration 5.10: Diagram sale from stock to internal consumers

5.14.2   **Prerequisite**

The planning and internal procurement must be run through as preliminary scenarios.

In the scope of the sale from stock to internal consignment following R/3 components are integrated used:

R/3 components	Functions
Sales	Customer order procedure
Shipment	Delivery procedure, Goods issue
Warehouse management	Materials transfer and release to/from stock
Quality assurance	Quality check for delivery and return
Transport	Transport disposition and dispatch
Foreign trade	Export control, reports to authorities
Invoicing	Invoice-, credit/debit memo-, rebate procedure
Information system	Planning, forecast, statistics

Table 5.8: R/3 components used in the sale from stock internal consumers

5.14.3   Procedure

Customer order

1. With a customer order a short-term agreement for the delivery of products to the other plant is determined. Herewith following central functions are available.

    - Calculation and pricing
    - Availability check
    - Demand transfer to the program planning
    - Dispatch determination
    - Export control

Dispatch

2. The dispatch procedure is initiated by the criteria of deliveries of orders due to dispatch. Herewith the availability situation and the dispatch determination validity are again checked.

3. Subsequently the goods withdrawal and the provision in the picking area are made in the scope of the picking.

4. The packaging of delivery positions ensues by allocation of dispatch elements, which can be a combination of materials, packing material or again dispatch elements.

5. The termination of the dispatch procedure is marked by the goods issue, which leads to a reduction of stock and actualization of the evaluation of the balance street accounts in the financial accounting.

Warehouse management

6. The combination with the warehouse management system (WM) takes place in the scope of the picking. Hereby it is to be distinguished between the picking in warehouses with fixed bin locations and chaotically organized warehouses (usage of transport orders).

Quality assurance

7. If a delivery note is generated for a product, the R/3 system automatically opens an inspection lot and provides the business documents necessary for the quality assurance. Nevertheless the inspection lot can also be manually generated.

If the product to be delivered is performed in batches and a delivery position is not to be satisfied in a single batch, the R/3 system can divide delivery position in several batches and generate a partial lot for every batch.

If it is determined that the quality assurance can follow after the delivery, the goods issue can be posted independently from the quality assurance.

If the inspection presents no evident defect, the feature results for the inspection and partial lot are entered.

8. Possible failures can be documented by failure date if the adjustment measures combined shall be managed in the R/3 system, a quality message/information is to be made

in the scope of the problem management and performance.

9. If the feature results to the partial lots are registered the usage decision for the partial lots and the total inspection lots is to be made. Otherwise a usage decision is made for the inspection lot.

Costs, having incurred in the scope of the result registration or the usage decisions, can be confirmed to the production planning at these times. The system automatically generates a quality certification if required.

10. If the quality check shows no result, the goods issue can be posted after the usage decision (has been made).

Transport

11. In the transport disposition the deliveries due to transport can be collected in transports. Essential disposition activities are the determination of dispatch dates, the determination of transport means and the route allocation.

12. The transport handling enables a monitoring of transport activities to be made after the disposition.

Foreign trade

Following functions are supported in the foreign trade procedure:

13. Export control on base of permits in order, delivery or goods issue.

14. Reports to authorities (e.g. INTRASTAT) on base of customer invoicing by an automatic reporting procedure.

Invoicing

15. The invoice creation is made on grounds of deliveries. It results in the procedure of postings to the corresponding financial accounting and cost accounting accounts, whereby here in general, only the costs are posted, not the revenues.

Information system

16. With the help of the information system planning, forecast and analysis, based on master and movement data referring to the business processes, are created.

5.15   Tank car procedure

5.15.1   Description

The sale from stock of tank cars to industrial consumers includes all processes necessary for the sales, in order to sell products from the stock. The particularity of bulk goods, which are sold in tank cars, is taken into consideration by a direct connection between the customer order and the tank car filling before the delivery.

In particular, these are the handling of customer orders with or without reference to quotations or outline agreements, the delivery and dispatch procedure as well as the invoicing. Further more the scenario includes processes for the procedure of customer complaints, subsequent compensations as rebates or commissions as well as returnable packages or empties (packaging and similar).

The scenario of the tank car procedure is graphically represented in illustration 5.11.

Single process steps of the various procedure forms of the sale from stock can be reduced to a minimum or even omitted. As e.g. the variant cash sale, which in general proceeds without the dispatch handling, since the customer takes the good directly.

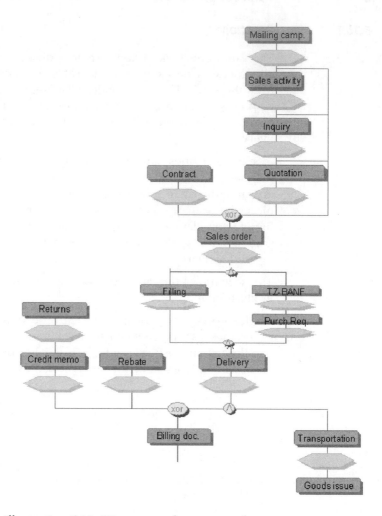

Illustration 5.11: Diagram tank car procedure

The scenario sale from stock via scheduling agreements deals exclusively with the operational procedure of products sold from stock. Consequently, there are serial products produced or picked on grounds of demand –respectively sales planning, are reordered demand – or forecast orientated and can directly be delivered to the customer (off plant). There is no direct relation between sales order and picking, the lead time of operational transaction is therefore relative short. Immaterial goods as services are as well dealt with at another space as all variants of a customer related production or final assembly of products.

The scenario is further more characterized by the categorization of the customer as industrial consumer in difference to the trading companies or final customers. The exception is here the typical procedure of the component industry, which is also described in a separate scenario.

The business processes described in the sections 5.1 to 5.15 are again built up by many single partial processes. In the following the most important partial processes shall be described.

5.16   Customer inquiry procedure

5.16.1   Definition

In general, the inquiry is placed by a customer to a customer caring department of company. The customer inquiries mostly refer to products, prices and delivery time respectively to the delivery stock. Customer target is, to gain information, which enables the purchase decision. The order cycle can start with the order or with a preliminary inquiry or with a quotation. Inquiry and quotation enable the entry storage of all important sale-related information used during the order procedure. Less complex inquires and quotations can be rapidly administered on the initial screen. Thy can be created anew or be copied. If the customer wishes for example a quotation, the data of a preliminary inquiry can be transferred to the quotation. Data stored in inquiry and quotation form the base for subsequent documents and, if required, even for the sales analysis. Sales employees can directly copy the data of a successful quotation into the contract.

5.16.2   Usage

The information demand of the inquiring customer is registered in an inquiry. Already in this phase of the business approximation all important sales related functions, which are necessary for the quotation and order handling can be administrated and stored. Further more, a broad base for the market processing shall be created with the inquiry procedure and storage. The inquiry is registered by the sale organization, which receives the customer inquiry.

5.16.3   Description

Customer inquiries can be received by mail, fax, letter, telephone, personal conversation and so on. Inquires are received by customer care centers at home and abroad. Inquiries can be related to different contents:

1. The customer has a technical problem:

    *   Does the company have a product, which is able to solve the customer problem?

    *   Do different quality steps exist?

    *   Are there alternatives to a competitive product?

    *   And so forth

2. The customer knows the product and asks for:

    *   Deliverable amount

    *   Delivery dates and prices

    *   Conditions

    *   Packaging means

    *   Transport possibilities

    *   And so forth

The inquiry must register the ordering party and the material. Further more, the system wants a requested delivery date, which is filled with the current date by the system. Additionally, the amount and the validity date should be registered, which gives information to which deadline the inquiry has to be handled.

With the inquiry registration it is checked, if the inquiring customer is already included in the customer master data file of the system. If not, he must be recorded in it (Here, the branching point into the customer master data procedure takes place).

In cases in which it is not clear whether a long-term customer relation ship can be built up, the definition of the inquiry customer as a CPD customer is possible. With no allocation of a customer number in the master record maintenance ensues. This results in a difficult reporting procedure or makes it even impossible.

During the inquiry registration, a system related check, referring the product is made. A text position can be registered in the inquiry instead of product numbers, if a material master record doesn't exist yet or it is not established which product corresponds to the customer requirements at the time of inquiry. If the customer inquires for a competitive product, which is not stored here in similar form, article alternatives must be determined (material finding). At this place, an export control is made, that is, it is checked, whether the material may be delivered to this customer at all.

With the registration of the inquiry, an automatic pricing and even a delivery capability check can be made according to the customer request. The registration of the inquiry serves merely for documentation. A recorded inquiry brings on following steps:

- A quotation shall be created for the inquiring customer: Quotation registration with reference to the underlying inquiry.

- A quotation should not be created for the customer. The customer in informed about the rejection of the inquiry. A reason for rejection should be deposited, for making clear that the inquiry is finished and possible data-recalls can be made about inquiries not leading to quotations.

- It must be checked, whether a quotation shall or can be generated. The affected centers are consulted with and the customer is informed about the loss of time.

5.17 Customer quotation procedure

5.17.1 Definition

The quotation is the seller's declaration of intention to a certain person or company to sell one or more goods or services at a certain time among certain conditions (prices, delivery dates and so forth). A quotation can be created with or without reference to a customer inquiry.

5.17.2 Usage

The quotation can be variously used:

Only with specification of a price

The current price is given to a present customer with reference to his usual order amount and conditions.

With specification of a price and order amount

The price is determined on grounds of a certain order amount, which shall be purchased in a certain time.

With specification to a price, order amount and a certain delivery date

The quotation is mostly referred to a concrete inquiry. The customer wants, beside of the price also the possible delivery date to be enlisted in the quotation.

Only with specification of amount

This quotation type is used for special sales (slow moving item, residual items). The good is offered to the customer and he is asked to place his order. One of the quotation types can be also applied for these special sales.

The registration of the quotation is in any case an optional function. It means additional work, but it enables the registration and storage of all important sales related information applied during the order procedure. The data stored in the quotation form the base for subsequent documents, and if required, even for a market survey. The sales employee can directly copy the data of a successful quotation in the order. With that it can be indefinitely referred to the quotation (with reference to the amount).

Further more, e.g. all quotations, which have been processed within the last 6 months, can be presented for the purpose of analysis, and those can be closer inspected which haven't lead to an order.

The quotation is also used for the support of the letter of credit procedure (quotation type 3). With an export order with the letter of credit, a quotation is first of all generated for the customer. With reference to this quotation a pro forma invoice is created for the customer and sent to. The order is only accepted and registered in the system, when the customer has sent a letter of credit for this invoice.

Only to this time the order amount is placed in a demand list, a delivery capability check made and a delivery date liable confirmed to the customer. The delivery date must lie in the validity time of the letter of credit.

The quotation is registered by the customer's sales organization responsibility and can subsequently be transmitted to the customer by fax, letter, telephone or EDI.

5.17.3   Description

Quotations are initiated both on grounds of customer inquiries and customer contacts (e.g. customer visits by the field service) and actively by the authorized company. Because of the possibilities mentioned above, a quotation creation and handling can be arranged in the system. Before this procedure step takes place, it has to be cleared, though, whether a quotation should be given to the customer or the inquiry has already been refused in this state (no product deliverer, production is overstrained, business relationship not required).

In case of a quotation refuse the customer is informed by telephone or in writing. Is a quotation to be prepared, then first it is to be checked whether the inquiring customer is already registered in the customer master record respectively authorized. Isn't that the case, the customer procedure must above all be branched in the customer processing.

In cases, in which it is not clear, whether a long-term customer relationship shall be built up, the definition of the inquiring customer as CPD (Conto pro Diverse; customer address and data are manually registered in the document entry) is possible.

Quotations can be referred to the underlying inquiries, be copied from already existing quotations or be completely newly registered. The quotation enables the written and liable reaction on the customer inquiry respectively information about new price conditions:

Required fields are:

- Customer (ordering party)
- Material
- Currency
- Amount unit
- Payment conditions
- Validity time
- Delivery plant
- Amount
- Delivery date
- Incoterms

Some of these data (which are written in italics) are controlled system related due to a consequent maintenance of the master data. But they can also be transferred.

The delivery date makes it possible, to state a certain day, a week, a month or a period.

An instruction is given within the registration of the quotation amount, if no complete load is registered. In this case a load surcharge is additionally registered in the pricing. This must be separately displayed on the written quotation confirmation to the customer.

During the quotation registration data checks are performed referring:

- The customer
- Material
- Product liability (export limits / product exclusion) and
- Delivery capability (only with delivery date inquired)

The credit limit check as well as the dangerous goods check are only performed during the order registration.

Manual *reservations* of batches can be made by the dealer, if required. An automatic reservation out of the quotation does not result.

Is the quotation registered, it can be transmitted to the customer.

If the customer accepts the quotation, an order registration with reference to the quotation is made. Herewith the quotation is finished.

If the customer does not accept quotation, the rejection reason should be displayed in the quotation, in order to mark quotation as finished.

5.18   Order processing

5.18.1   Definition

An order is a two-paged declaration of intention between seller and customer, which aims at the production of the specifications agreed respectively at the acceptance. With this declaration of intention the customer is obliged to accept the ordered goods and/or services to a fixed price and delivery date from the seller.

An order can also refer to a quotation. Thus, this quotation is liable, if the validity duration and the contract conditions are met.

5.18.2   **Usage/Task**

The most different customer requirements shall be registered and administrated in a system, so that all participating sales organizations are enabled, to inform their selves at any time about the current level of customer order.

The order processing pursues the use of customer requirements formulated, respectively transmitted, within an IT supported system, on different ways:

- To register

- To separate in single, logical subsequent working steps

- To supervise and control and

- To inform all positions participating within the order procedure,

in order to meet all customer requirements.

In the following, the standard order processing is described. The peculiarities within the special order types and some business transactions are explained in the subsequent chapters.

5.18.3   **Description**

The customer places an order in three ways:

- On base of an existing, valid quotation

- With reference on a contract, order or invoice,

- Without reference

The marketing decides, whether an order is accepted. If the order has been based upon a quotation and it is up to the moment temporally valid (validity date from – to), provided that the order data correspond with the quotation content, the order has to be accepted. In case of differing data between quotation and order, the customer has to be contacted.

After an order acceptance it first of all has to be checked whether the ordering customer is recorded in the customer master data respectively whether an order acceptance is still authorized (e.g. exclusion on grounds of an overdraft of the credit limit and so forth). Before a further order processing takes place, in the negative case, it has to be branched in the process "customer procedure".

In cases, in which it is *not* evident whether a long-term customer relationship can be built up, the definition of the inquiring customer as CPD-Customer is possible.

The order is generally registered in the SAP system. There the order is administrated beyond the total order handling procedure. Ahead of the order record the recorder has to decide, whether he wants to refer to another order or a former invoice or to a quotation fixed in the system (generating of an order referring to a reference document or but to record the order anew). The system automatically gives an instruction information to the corresponding documents to which it could be referred, if an order with material is registered for the customer (without reference), for which, according to the customer a valid quotation or an authorized contract does still exist (in the system).

If a quotation or a contract is not any longer timely valid, it cannot be used as reference document anymore. Shall the quotation nevertheless serve as reference for the order to be registered, either the validity time of the order must be changed or a new order created. If this does not take place, the header and position data necessary must be manually recorded in the document.

Every order is multi-position capable, that means 1-n position pro order can be registered.

All header data of the documents in hand are taken over into the order during the reference to existing documents. The quotation can be taken over totally (all document positions) or partially (only special positions) into the order. The document data copied must be checked. The data can be supplemented or changed.

Following fields are required fields in any order header:

- Customer (ordering party)
- Invoice receiver
- Payment condition
- Dispatch condition
- Goods receiver
- Payer
- Currency

For the order positions following fields are required:

- Material
- Single price
- Price quantity unit
- Delivery date
- Amount
- Sales quantity unit
- Dispatch position/route
- Incoterm

By consequent maintenance of the master data (e.g. customer, product, customer-product-relation, price, and so forth) a major part of these data can be controlled system orientated (a super-scription/supplementation is possible in the individual case).

The batch, which corresponds to the customer requirements, is in general not known to the marketing employee at the time of the order registration. Only in exceptional cases the order is recorded with batches (e.g. residual item) by the customer care manager. An automatic batch allocation doesn't take place. The batch is only registered by the disposition in the delivery document.

The delivery place is also registered only by the disposition with the creation of the delivery.

On header and position level texts for the internal respectively external information sent-ahead can be manually or with a text key (distribution to standard text) recorded. In the same way the record of export data is possible. If a position cannot be completely delivered to a delivery date and if partial deliveries are agreed upon with the customer, a position subdivision is registered per delivery date of this position (in connection with the partial delivery amount).

If e.g. a position with 100 pieces can only be delivered at four different delivery dates, then four schedule lines are generated for this position, to which the corresponding quantity and the delivery date are displayed. Every schedule line is in the further course of the business process separately dealt with.

The order data are basically subjected to a syntax check (formal correctness) during the registration. Incorrect order data are to be immediately corrected by the registration.

During the order record additionally following checks are initiated in the header respectively position level:

- Delivery capability check (availability of the product at the required delivery date)

- Credit limit check

- Dangerous goods

- Export control check

Further on the functions pricing, passing on requirements, dispatch schedule, dispatch position and route calculation are performed.

With the registration of the order quantity a memo is given, if no complete load is registered. In this case even the pricing achieves a load surcharge. This must be separately displayed on the written order acceptance (and further on in the invoice).

If content system checks result in complaints within the registration of an order, it is to be clarified, whether the other data can be changed on position level. In some cases contact has to be established with other positions (production, customer) for clarification.

Then the possibility exists, to secure the negatively checked order up to the final clarification and to block those functions following the order procedure. The sales document is the base for different subsequent functions as e.g. the delivery handling and invoicing, though the subsequent functions can only be performed, if the data in the sales document are complete. For the guarantee of this completeness, the system registers in an incompleteness protocol all data still missing and points out to the register that further entries have to follow. The possibility of creating a list of all incomplete documents is offered to the register. The decision about the further procedure lies upon him. Changed order data on the other hand pass through an outlined check routine.

The customer is informed about a cancellation in case of final negative checks, in case of orders accepted the customer achieves an *order acknowledgement* only if the required. This information can proceed in writing, by telephone, but also by EDI.

Order entry

Orders to be entered can be on one hand referred to a quotation or an order already done or registered completely anew. The order receipt can be achieved by post, telephone, fax or telex.

With the customer order acceptance an order is accepted by the marketing. The system automatically proposes data from the corresponding master records:

- Sale, dispatch, pricing and invoice data are proposed from the customer master record of the ordering party. Moreover, the system copies customer specific master data to texts, partners and contact person at the customer.

- The system automatically proposes data for every material of the order out of the corresponding material master records as for example data for pricing, dispatch scheduling and volume determination.

Data proposed by the system can serve as basis for the order, but if required, can also be manually changed or added anew.

Order procedure

With the order procedure following functions are performed.

- Pricing
- Availability check vs. planned independent requirements, as long as such a check according the material master record is demanded.
- Requirement transfer
- Dispatch schedule
- Dispatch center and route handling
- Credit limit check

In addition, a dangerous goods check is manually made, if one of the positions includes a dangerous good. Further more an export inspection is affected. By that it is controlled if a general embargo is laid on the country of destination, if the goods re-

ceiver is registered in a blocking list or if a good is subject of authorization.

Batch finding

With the order entry in general, neither batches are specified nor
an automatic batch finding takes place. The responsibility for the
batch specification lies in the disposition, where the good is disposed of (in the delivery document). In special business transactions (e.g. NT-good, residual item) the batch is already given
along in the order.

Information about packaging of goods are also first of all registered in the delivery document.

With the entry of an order the good's retirement is noted correspondingly to the delivery data registered. The good is not reserved in sense of the MM. But by checking and billing vs.
planned independent requirement only a limited order stock is
authorized. If an order with a higher delivery priority gets into
the system, the employee him/herself must raise the delivery
date already registered in the order, in order to release goods for
a new order or increase the planned independent requirement in
accordance with the product.

Delivery processing

The 'delivery' processing follows the 'order procedure'. Here, the
transport volume disposition (disposition and procurement of the
transport volumes necessary), the dispatch itself (creation of the
delivery document picking, provision of goods at the loading
point), the document procedure, the creation of the customer
and dispatch data, the creation of papers and caution marks necessary for dangerous goods, the transport procedure, the loading
and weighing as well as the goods issue processing are performed. Herewith, all steps necessary for the dispatch of a good
to a customer are processed deadline related (starting from the
customer desired date the deadline invoice is initiated).

With the creation of the system delivery note, free stock volumes
are related to concrete order positions (automatic stock decrease
in site and storage location and posting in 'stock for dispatch')
and proofed in a picking list. The warehouse clerk arranges the
shipment to the customer by means of this information. After the
packaging and marking procedures and the shipment to the
transport mean, the goods issue is posted in the system, whereby
the exact amount is posted from 'stock to dispatch', with loose

good only after the weighing at the gate and a corresponding confirmation.

Topical to the goods issue, the 'invoicing' of the delivery and the transfer of invoice data to the customer financial accounting for posting and supervision of the collection take place.

5.19   Delivery procedure

5.19.1   Definition

The delivery is the physical provision of a customer with the goods demanded in the right quality, to the right time and to the right reception point. In general, deliveries are generated with reference to a customer order. Hereby the delivery is a document, generated in the R/3 system. Whereas a delivery note is a dispatch paper generated on grounds of the delivery document.

5.19.2   Usage/Task

The procedure of deliveries in the dispatch follow the sales order procedure. In the order it has agreed upon the delivery of a single or several articles to a certain time and price. To meet this order, the good must be provided in time and the dispatch planned and started time related. The procedure progress must be controlled unless the good leaves the site. The goods issue terminates the delivery referring the dispatch. The procedure can now be invoiced.

The central document in the dispatch is the delivery. It serves as initiation of all dispatch relevant activities as picking, confirmation, packing (in the sense of SD), print of dispatch papers and goods issue. The delivery is the basis for the transport and invoicing. If in the delivery an order is referred to, the element dispatch data are taken over into the delivery document.

5.19.3   Description

The dispatch activities in the sales are initiated with the creation of a delivery. Since all information necessary for the dispatch procedure have been taken over from the customer order respectively master records, a delivery can be in general created in the system without manual expense. Following activities are proceeded by the system with the delivery creation:

- Order and material control, in order to determine the possibility of a delivery (delivery blocking, incompleteness and so forth.)

- Determination of materials and amount/qualities due to dispatch

- Availability check in the customer order, an additional availability in the delivery can be made.

- Consideration of possible partial deliveries

- Route determination

- Allocation of the picking storage point

The processes picking and packing follow the delivery procedure.

The picking transaction includes the withdrawal of the good off the warehouse in the field of picking, where the good is prepared for dispatch. The efficient picking is absolutely necessary in order to guarantee an optimal customer care management.

After the entire picking the good is packed and subsequently the goods issue procedure starts. The business transaction is finished on behalf of the dispatch, by the good leaving the company. This is presented in the SD by the posting of the goods issue for a delivery. This results in following consequences:

- The stock reduction is presented in the financial accounting and in the general ledger by evaluation and update of the corresponding stock accounts.

- The material demand is reduced for the delivery.

- The delivery status is adjusted.

5.20   Invoicing

5.20.1   Definition

An invoice is the notification to the service receiver about the remuneration becoming due on base of the contract.

5.20.2   Usage/Task

With the invoice the customer is charged on grounds of deliveries and services created out of a customer order. It serves the realization of a sale. Further more, credit and debit memos, which are delivery related and neutral, are presented in the invoice. The invoice is created in the name of the company to which the customer is related.

5.20.3   Description

Invoicing is the last planned activity of the sales. The invoicing generally initiates the system related posted goods issue document and the physical leaving of the good off the delivery plant.

After the system related invoice posting is made, order delivery or invoice data cannot be charged or cancelled any more. Alterations are only possible due to credit or debit memos, cancellations merely due to reversals. The reversal of invoices can be necessary e.g. after failures during the invoice creation or posting (account determination).

Following invoice types are required for the procedure of the different business transactions in the sales:

- Delivery-/service related invoice quotation
- Pro forma invoice for the quotation (letter of credit requirement/ without amount reservation)
- Pro forma invoice for the order
- Pro forma invoice for the delivery (good accompanying invoice)
- Reversal invoice
- Credit memo
- Return credit memo
- Reversal credit memo
- Debit memo

Invoices can be created in three different ways:

1. The documents to be invoiced can be explicitly specified by placing the numbers of these documents or searching these with the help of a match code.

2. The single documents to be invoiced are not to be explicitly specified within the procedure of the invoice stock. On ground of the stated selection criteria the system collects documents, which have to be invoiced.

3. The invoicing can run in the background.

5.21   Contract procedure

5.21.1   Definition

A contract is an agreement (outline agreement) with a customer upon an acceptance of certain products in a certain quantity (target quantity) and certain period. The contract includes fundamental quantity and price information, but no determination of the delivery date and amount (also look at delivery plan).

5.21.2   Usage/Task

Contracts are made with customers who frequently order certain products. With the termination of a contract following advantages occur for customer and deliverer:

Customer:

In general fix prices and conditions during contract period

Deliverer:

Knowledge about the sales volume of the products defined in the contract during the whole period.

For a contract preparation it is checked, whether the customer already has a master record in the customer database respectively an existent customer database is authorized without restrictions for the sales procedure (no permanent credit limit exceeding and so forth).

If the customer master record is not yet available, a customer processing is first of all branched.

A given quotation is often the basis for a contract. Therefore it is to be checked, in what way a temporal valid quotation exists in the system. The leader data of the quotation are liable for the contract and are completely copied into the contract document. The contract positions can be copied out of the valid contract, modified and supplemented, otherwise the position can be completely manually added.

With the generation of a contract it is also possible, to refer to an order, invoice or another contract of the same customer.

If no quotation exists respectively the validity period is terminated and no reference document is available, the contract data are to be manually registered (the directing of information from the master records).

Following fields must be given along for the creation of a contract:

- Customer number of the ordering party
- Validity period of the contract
- Material number(s)
- Target quantity for the materials (target quantities aimed at)

As for all other order types also, the possibility exists, to individually register further data for the header or the positions by the corresponding menu entry and the matching pictures.

All header and position data of the contract are subject to a system related inspection in reference to syntax and content correctness. Possible failures have to be solved. Is no immediate solution possible, the current (incorrect) contract is to be stored up to the final clarification.

An automatic pricing takes place with the registration of the positions.

The at last correct contract is stored in the system. A contract agreement can be sent to the customer, if required.

5.22 Release order procedure

5.22.1 Definition

A release order is a subset of a contract, which is released to a defined delivery period. The release order is opened with reference to an existing valid contract in the system. The data agreed upon in the contract are transferred into the release order.

5.22.2 Usage/Task

The contract doesn't include any schedule with delivery data and therefore must be released by an order. The delivery releases with quantity and date are made by the customer.

5.22.3 Description

The customer care manager accepts the contract release and first of all checks, if the contract agreed is still valid. If this is not the case, the customer is to be informed and the further procedure negotiated.

With the validity of the contract a release order is initiated in the system, whereby the field contents of the contract are copied in the order release and if necessary are supplemented by header text and export data.

The order header created in that way is system related checked for syntactical and content correctness. Failures are immediately to be corrected. The quantity can be accepted or also be changed.

The delivery dates and if required texts and export data agreed upon can be manually maintained. With contracts handled in batches the entry of batch numbers is optional (with the delivery note handling at the latest, the batch numbers for products handled in batches are to be entered respectively are automatically distributed by the system).

In the system, checks referring

- Product authorization
- Delivery capability
- Credit limit
- Dangerous goods
- Export

are made for all products listed in the order release. System related complaints are to be clarified and corrected. The correct order release is stored in the system. An order agreement is generated and transmitted to the customer, if required (explicit customer demand). For those release orders the delivery handling and invoicing takes place as with any other standard order. In the course of the release order procedure document alteration, or reversals (by customer or deliverer) can become necessary. Every update in the order release goes through the system checks mentioned above and reported to the customer by an update document. Generally valid updates are also to be made in the contract, for the future. Every quantity released automatically decreases the contract quantity (quantity update). Total contract quantity updates by single releases initiate a system warning, that is, a quantity exceeding is possible, but is reported by the system, in order to reduce the quantity of the last partial delivery (compliance with the contract quantity agreed upon).

5.23. Delivery procedure free of charge/subsequent delivery

5.23.1 Definition

A delivery free of charge respectively a subsequent delivery free of charge is the consignment and transfer of goods to a customer without calculation.

5.23.2 Usage/Task

A delivery free of charge is registered e.g. in order to send a model to a customer. A subsequent delivery free of charge follows, if e.g. on grounds of a reversal to an order material is to be subsequently delivered (e.g. with damaged or missing goods). In this case it is referred to the original customer order. In general, subsequent deliveries free of charge are rather seldom. Normally, the customer gets a credit memo. Only on request of the customer (merely with goods packed) a subsequent delivery free of charge is initiated.

5.23.3 Description

For the registration of a delivery / subsequent delivery free of charge there are the order types KL or KN.

A delivery free of charge is in general not generated with reference to another document. It is the first document within a business transaction. Thus, a reference document can be applied as copy help.

The entry of an order reason is obliged. If the register forgets the entry, he/she is informed by the system about the incompleteness of the document.

With the enter of a subsequent delivery free of charge, the entered by is always taking reference to a preliminary business transaction (order or invoice or return order). Though it is not necessary, to choose the function 'create with reference' since after the entry and pressing the ENTER-button a dialogue window automatically displays, in which the number of the reference document is required, the system automatically places a delivery block, if a delivery free of charge or a subsequent delivery free of charge is created. This block is manually removed, if the transaction has been controlled and released. The block indicator is placed on the header picture with commercial data. Only when the block indicator is removed, a delivery can be generated. If the delivery/subsequent delivery free of charge has been rejected, the reason of rejection must be recorded in the order in order to mark the order as finished.

5.24.   Credit-/Debit memo procedure

5.24.1   Definition

Hereby it is referred to subsequent advance payments respectively surcharges for the customer invoice with following profits or losses.

5.24.2   Usage/Task

Customer invoices cannot be changed after a system related posting. If an alteration reason occurs, a credit or debit memo becomes necessary. Such reasons are surplus or reduced power on grounds of quantity corrections, price conditions, distribution of rebates and complaints of quality defects as well as returns.

5.24.4   Description

The credit-/debit memo request is the reference document for the creation of a credit-/debit memo and is created on grounds of a credit memo by the customer care manager. It is blocked for inspection. Is the request legal, it is released and a credit memo created.

This credit-/debit memo request is an order type for it's own. It is entered without reference or with reference to the corresponding invoice or an order (e.g. return delivery) in the system with a key for the credit-/debit memo reason.

There are different reasons for credit-/debit memos:

- Quantity or price failures in the invoice
- Repayments by agreements at the end of the month (or also during the month)
- Rebates surcharge by termination of a period (in general quarterly as agreed upon)
- Rebate distribution – reversal of accruals for rebates
- Customer complaints on grounds of quality defects or mis-dilivered goods

- Return delivery because of quality defect of misdelivered goods.

- Defect credit-/debit memos

Hereby the catalog of credit memo reason is to be checked, whether all reasons have to be transferred into the system.

Credit-/debit memo

The header and position data are transferred to the credit-/debit memo request with the entry referring to a reference document. Subsequently the header and position data are to be manually adjusted, if required. The credit/debit memo quantity consists of the order quantity minus a possible already used quantity. It is to be changed, if required. The pricing is defined in the standard version of the R/3 system in that way, that the price parts can be transferred unchanged with automatically or manually entered surcharges and advanced payments from the reference documents. The statement of goods and services is of importance, however since the credit /debit memo refers to former transactions and the tax is determined according the tax level at the time of the goods and services (delivery and invoice).

Thus, the transaction has been performed, checked and provided with a reason. After the creation, a block is automatically placed with the credit-/debit memo request, which can only be released by a further person.

After the decision made and the release followed, the credit-/debit memo can be initiated. Whether a paper storage with handwriting authorization and signature is necessary, is to be decided by the revision.

With a rejection (no release of position) the rejection reason can be entered with the help of a key per position, which is stored in the system. These positions are taken over in the credit-/debit memo as statistic positions without consideration.

The creation and further procedure of credit-/debit memos corresponds to the invoice procedure. Here can also a credit-/debit memo be created out of several reference documents. Credit-/debit memos can be explicitly entered or an invoice storage be proceeded. Credit-/debit memos can merely be generated with reference to reference document (e.g. a credit memo is reported with reference to a credit memo request).

The data of the reference document are automatically transferred in the credit-/debit memo and are not to be changed any more. With the storage of the document or after the creation of the document in the system it is either printed or sent to the customer as a master record and transmitted to the account.

Rebate credit memo requirement

A rebate credit memo requirement is recorded, in order to invoice the rebate agreement in the SAP R/3 system. This serves for the explosion of rebate provisions or for the handling of payments with provisions. It always refers to an agreement. Further descriptions follow in 'Rebates and other procedures dependent on provisions'.

After the creation of the rebate credit memo requirement the person must ratify it in the system (see above). On grounds of this a credit memo is automatically generated and the provision exploded.

5.25   Credit limit check

5.25.1   Definition

The **credit limit** is a concessive maximum for the guarantee of a customer credit for every customer.

5.25.2   Usage/Task

The determination of the credit limit and its maintenance in the customer record lies in the responsibility of a company's finances. The credit manager determines the customer specific credit limit in accordance with the marketing. This amount is registered and maintained in the credit management picture of the single customer master record. Single credit limits can be given for every credit control area and the credit of every customer can be controlled, also by giving a central credit limit for all credit control areas to which the customer is allocated. The whole limits on the level of the credit area are not to exceed the total limit of the credit control area. The credit limits on the level of the control area are checked during the order procedure.

In the scope of the sales procedure the customer credit limit is system related checked within the order procedure and dispatch procedure. Hereby the limits put opposite to the open de-

Vieweg Verlag

Zukunft seit 1786

Vorsprung im IT-Business

Ich bestelle zur Lieferung über meine Buchhandlung:

Expl.	Autor und Titel	Preis

Besuchen Sie uns im Internet

Antwort

Vieweg Verlag
Buchleser-Service / LH
Abraham-Lincoln-Str. 46

65189 Wiesbaden

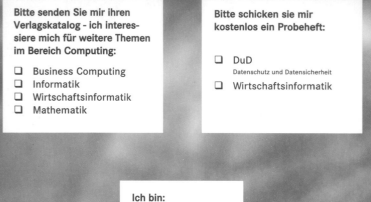

Bitte senden Sie mir ihren Verlagskatalog - ich interessiere mich für weitere Themen im Bereich Computing:

❏ Business Computing
❏ Informatik
❏ Wirtschaftsinformatik
❏ Mathematik

Bitte schicken sie mir kostenlos ein Probeheft:

❏ DuD
  Datenschutz und Datensicherheit
❏ Wirtschaftsinformatik

Ich bin:

❏ Dozent/in
❏ Student/in
❏ Praktiker/in

Bitte in Druckschrift ausfüllen. Herzlichen Dank!

Firma

Abteilung

Vorname

Name/Titel

Straße/Nr.

PLZ/Ort

Telefon*

Fax*

Geburtsjahr*

Branche*

Funktion im Unternehmen*

Anzahl der Mitarbeiter *

Mein Spezialgebiet*

* Diese Angaben helfen uns, Ihnen einen passgenauen Service zu bieten.

Wir speichern Ihre Adresse, Ihre Interessengebiete unter Beachtung des Datenschutzgesetzes.

322 01 200

vieweg

mand/orders by taking in consideration the customer individual payment target (dynamic view).

Following functions can be blocked by the credit status:

- Generating of material reservations

- Generating of order requirements

- Generating of production and planning orders

- Generating of indices due to dispatch

- Print of order acceptance

- Generating of deliveries

In that way, an eventual impending dept loss can be countered already in the starting of the sales procedure (reversal of customer order charges of the payment target to advance payments or delivery block, unless outstanding customer demands/debts are totally or partially settled.

Special automatic credit checks can be defined, which meet the specific demands of the credit management areas. The inspections can be undertaken to different dates with the order procedure – from the order receipt up to the delivery.

In the dispatch following functions can be blocked with the credit status:

- Picking

- Packing

- Posting goods issue

- Printing of delivery notes

5.25.3 Description

In the credit limit check process the single document position (quotation, order, delivery note) are adjusted with to the customer individual credit limit with its value and with consideration of the payment conditions. The single customers are allocated to certain risk classes. The system's reactions in the single sales documents to an exceeding of the credit limit are controlled by these risk classes. The definition of the risk classes and the allocation of the customer to those is also made in the SAP-menus respectively views, which lie in the responsibility of the finances.

Credit limit evaluations

The SAP standard system contains different reports for the supervision of the credit management.

- Report RFDKLI40 (overview credit limit)

- Report RVKRED06 lists all document blocks out of the credit view

- Report RVKRED77 serves for the rebuild up of the open credit, delivery and invoice values

- Report RVKRED08 checks all sales documents anew, which are part of the horizon of the dynamic credit check.

Credit limit check types:

- Static credit check (open orders, deliveries, invoices, open position in comparison to a fixed credit outline)

- Dynamic credit check up to the credit horizon (static check plus open order value within a specific period)

- Maximal reference value (order or delivery value)

- Date of the next internal check for the automatic check

- Overdue open items in reference to customer balance

- Oldest open items limited to a certain quantity of days

- User checks (possible with special user exists)

5.26   Standard order with advanced payment

5.26.1   Definition

A standard order with advanced payment is a common standard order, with which the customer has to pay the good before its delivery in advance.

5.26.2   Usage/Task

The company can insist on advanced payment by the customer if an order is proceeded with a new customer or with a customer, whose' credit limit is already exceeded or not large enough to cover the order value.

5.26.3   ## Description

There business transactions are registered as normal **standard orders**. In the order, however, a delivery block is set on header level, which is only reactivated, if the customer payment has been achieved. The customers are allocated to certain risk classes by the credit management. Among others, risk classes control whether a delivery block is automatically set by the system. New customers are allocated to the risk class 'N'. This means, that they are automatically delivery blocked from a maximal document value of 0,51 EURO. Only if the credit worthiness of the customer is checked an advanced payment is no longer necessary, he is allocated to another risk class. This delivery block must be manually set with customers of risk class 'A' (less risk). With the customers of the other risk classes the delivery block is automatically set (since the credit limit is not sufficient).

In all cases it is to be duplicated in the system, on what grounds the delivery block has been set.

After saving the order, a report is made to the customer finance (by mail), that the customer payment is expected.

The customer achieves a written message (order confirmation or short telefax) or a message by phone about the range of the invoice. The creation of a pro forma invoice is not necessary (the invoice amount is registered in the order confirmation). The payment receipt of the customer is posted in the FI and must be reported to the customer care manager (by mail).

After that, the order is treated like a common standard order. If the transaction is invoiced, the allocation of the amount is made in module FI. Moreover, possible discrepancies are to be corrected (return of the surplus to the customer respectively subsequent claims). This lies in the responsibilities of the customer accounting.

The good is provided with the order entry towards the planned independent requirements. Towards an existing stock the order amount is only provided after the release of the delivery, that is after the payment receipt.

5.27   Model delivery

5.27.1   Definition

Model deliveries are understood as the dispatch of products to the customer testing or introduction of the product.

5.27.2   Usage/Task

If the customer wants to test a new product or batch, to try an already applied product in a new installation, to change to another product, or the company develops together with the customer a fitting product, a sampling takes place. A test often requires small, but also larger amounts. The customer achieves a delivery independency with a business transaction, an expected order quantity with a successful test, the importance of a customer and the transport costs combined with the model delivery either free of charge, with a test rebate or to a normal product price but without small quantity surcharge.

5.27.3   Description

The order type *delivery free of charge* is applied for a model delivery free of charge. In the further course the normal delivery procedure is used (possible partial loads). An invoicing follows in this case with the note 'nominal value delivery free of charge'. In the order header the order reason 'sampling' is given with (for account finding CO).

If a model delivery free of charge and a normal delivery are combined in an order, a standard order is created and the model delivery is recorded as position free of charge (natural rebate). Here, also the order reason 'sampling' can be given with in the order header, but it is to be aware of, that this order reason only refers to the position free of charge. Therefore it is recommended, not to combine invoiced positions and positions free of charge in one order.

With every delivery unbilled, the release for the dispatch must be made by an authorized person.

For model deliveries billed, the normal customer order (*standard order*) is applied, and a test rebate is made or a small quantity surcharge released.

6   Master data

6.1   Definition

The operational basic data of a company and its' quality are the fundamental basis for a successful procedure of business processes. Herewith it is important, that the master data optimally support a single transaction (e.g. customer order), an overlapping view (e.g. company controlling) and the organizational structure of a company in the same way. The master data in module SD meet these requirements. On the one hand, e.g. customer- and material master data are extraordinarily outlined for the requirements in the sales; on the other hand the requirements of the financial accounting or material management are realized in the same master data, by the unique master data concept. Beyond that, in the master data the possibility is reflected, to adjust the system to organizational conditions. A comprehensive authorization concept serves for the protection of the master data.

With the operational standard software package SAP R/3 the basic data are recorded as so called 'master data'. The master data in the R/3 system are stored in a central module and form the basis for all business processes in the sales. These are information about business partners, material prices, surcharges and advance payments and data e.g. routes. The master data validity can be limited on partial fields of the company organizations, as e.g. on sales fields.

6.2   Master data in module SD

The master data in the R/3 system are, as already mentioned, stored in a central module and form the basis for all sales business processes. Essentially, information referring following fields, are included in the master data.

• Business partner

• Conditions

• Customer-material information

• Materials (products and services)

- Collections

6.2.1   Business partners

- In the module SD business partners are natural and juristic persons, with whom the company makes business. The differentiation of these business partners results from so called 'partner roles'. These are:

- Customers ordering goods or services. These customer data are of importance for the sales as well as for accounting. Because of this, both departments refer to the same customer master record. In the customer master record, also called debtors master record, three different types of data are filed: General data, company code data and sales specific data.

- Contact persons at the customer, as e.g. buyer or decision maker. The contact person data are also stored in the customer master record. This implies the name of the contract partner, the telephone number and department as well as further data.

- Deliverers, who deliver goods or services for the single business processes, including the forwarder agents.

- One's own employees, participating with the business transaction

6.2.2   Materials

In the module SD the term material includes both products and services. All data, necessary for the administrator, the inventory management and the sale of material stocks, are stored in the material master record. Material master records must meet the requirements of many users. The sales procedure is essentially based on the use of these master records. In the system SD the material master record is used with the inquiry, quotation and customer order processing as well as in the dispatch and invoicing. The material master data maintenance can centrally or decentrally ensue in the specialist areas. Herewith different data can be necessary.

6.2.3   Customer-material-information

In order to relieve the communication with the customer, customer specific requirements for the sales or deliveries of certain materials can be stored in the customer-material information record. Following data are possible:

- Customer individual material numbers and items
- Special data for delivery and delivery tolerances
- Texts

The data in the customer-material-information have priority over the general data of the customer respectively material master record. With the registration of a customer order a customer material information record is maintained, the data out of that are considered.

6.2.4 Conditions

With the pricing in SD it is possible, to perform both simple price structures and also complex, on each other depending, price elements. Target of the pricing is, to determine prices, surcharges and advance payments, freight and taxes (conditions) for a business process and additionally to allow a systematic, manual influence of these conditions. These conditions data are stored in the condition records and can be established on header or position level. The pricing described in the further course of this book ensues on basis of already defined price elements. These price elements are variants, with which the pricing can be adjusted to company specific requirements. Examples of the marketable prices in practice, which can be stored and maintained in the SD-system, are:

- Price lists
- Material price
- Customer individual price and so forth

Surcharges and advance payment can also be created, if required. As examples should be mentioned here:

Surcharges and advance payments

- To the customer
- To material
- To price groups
- To material groups
- To customer/material groups

- To customer/material
- To price groups/material groups
- Others

Essentially, the taxes are surcharges. Sale taxes also can be applied by the price elements in the module SD. These price elements for following taxes:

- Value added tax
- Taxes in dependence on the country code (USA)
- Sales taxes in dependence on the city code (USA) and so forth

With the automatic tax calculation in module SD it is taken into account:

- Whether it is a national or international transaction
- The tax classification of the goods receiver
- The tax classification of the material

Further more, other sales taxes not defined in the standard can be added. In this case the tax amount is determined in the customer order per position.

Referring the freight conditions it is distinguished between Incoterm 1, Incoterm 2 and others.

6.2.5   Collections

Often occurring material combinations and usual delivery quantities are stored in collections. With the order entry in the system SD a position proposal can follow then, with reference to a collection. Thus the order registration can be performed more efficiently, and that also reduces the failure frequency. It is also possible to change in the customer order positions proposed in the collection.

6.3   Efficiency features of master data

The master data in R/3 enable a fast/rapid company specific adjustment and offer a comfortable handling. Customer data and also data referring to other business partners can be defined area affectedly. In addition to it, the master data offer following further efficiency features:

- Predefined views on the material master with the possibility of a central and application specific performance.

- Flexible data structures for a comprehensive pricing and tax calculation in the sales.

- Creation and maintenance of the master data with the help of an entry mask, which can be adjusted to the individual requirements.

Material master records must, like customer master records, also meet the demands of many users. The sales processing is essentially based on the usage of these master records. The material master record in the system SD is used in the inquiry-, quotation- and customer order procedure as well as in the dispatch and invoicing. With that particular different data can be necessary. For this reason, the material and customer master record is in the following exactly inspected.

6.4   Material master

6.4.1   Introduction / Delimitation

In the following especially the sales and distribution views and professional requirements of the sales and distribution on the material master are documented.

6.4.2   Definition

Products and services are summarized by the term 'material', in the SAP R/3 system. All information necessary for the administration of a material and its' stocks as well as its usage, are maintained in the so-called 'material master record'. The material master records must meet, like the customer master records, the different requirements of many users. The sales procedure is essentially based on the usage of these master records, in the scope of the integrated standard software SAP R/3. Within the module SD the material master record is used in the inquiry, quotation and customer order procedure as well as in the dispatch and invoicing. The material master, by the way is applied by all components of the SAP-logistics system.

6.4.3   Use/Function

The material master is jointly used by many departments of the company and is therefore divided in different/various views. A material master can be created in a central (for all views) or decentral way in the departments. For the sales the sales view 1 and 2 with sales/plant data view and the sales text are to be created. The foreign trade view is part of the sales view and also of the purchasing view. Is only one of the sales views generated, this takes automatically place with the other sales views and the foreign trade view for the material. The automatically created views contain no data.

The sales views in the material master are hereby applied in dependency on the separate purchasing organization and the sales path for a sales department. On grounds of a connection between a material and a certain sales path it is possible, to sell the material to different conditions along several sales paths.

6.4.4   Material types

Material types in the system comprise trading goods, semi products, finished products, services and packaging materials. Due to the combination of those materials to material goods, e.g. screen sequences are defined within the transaction procedure as well as selection functions and number assignment. Therefore, different data are relevant in dependency on the material type. Following material types are important for the sales:

Finished product – FERT

Finished products comprise all materials produced in one's own company. Among others, production, disposition, and dispatch types can be maintained in the material master data.

Semi-finished product – HALB

All materials not yet finished for the production process. HALB-materials can be externally produced or sold to customer as well.

Trading good – HAWA

Trading goods, which are not produced, but purchased and resold by one's own company. In the material master records, purchase and sales data are filed among others.

Packaging – VERP

The material type combines all materials necessary fort he packaging. Among others e.g. card boards, boxes.

Product groups - PROD

Collection of material and other product groups

Services – DIEN

Services are immaterial goods, which particularly differ from other materials in that way, that the production and the consume take place simultaneously. Services are not transportable and are not kept in stock. They comprise among others, industrial services, transport services, banking and insurance company services and so forth.

6.4.5   Sector

The sector has like the material type an important control function and determines e.g. the selection of data fields on screens or the screen sequences.

6.4.6   Division in the material master

The division in the material master can be used as organization criterion. Prerequisite is though, that there are several divisions. The system warns with the message 'Proposal by the customer master not possible without a material division', if a customer-material-info-record is created for material without material division. Despite of this message all data are correctly taken over of the customer-material-info in the order.

6.4.7   Number assignment / Representation of the material number

The material number can be system-internally or externally assigned for all material types. The material number is 8-digit. For the representation of the material number no edit format is used.

6.4.8   Handling in batches

The batch maintenance is a component of the R/3 system, which is used in all fields of logistics. This it is mainly used by chemical industry, pharmacy industry and food industry, in order to keep separately in stock partial amounts of single materials.

A batch is the quantity of a material, which is produced in a manufacturing path and therefore performs a certain unit with a certain specification. As such a unit, the batch is provided with special physical, technical and chemical attributes, on grounds of which it can be described.

Within the management of the single batches following functions are available:

- Batch finding
- Batch status management
- Batch monitoring

The responsibility of the batch reservation in a company lies in the MRP. Batch specifications are generally not made in an order. The features of the different batches are often the production determined by the quality assurance. So called inspection lots are automatically built. Corresponding the controls, a usage decision is established, which determines whether a good is posted in a free available stock, which quality code the batch receives or whether it is posted to an other material because it is not type related. The quality code serves most of all as facilitation for the allocation of bates to customer orders.

6.4.9   Not type-related good

Not type related goods are goods, which do not relate to the specifications required. Depending on the usage procedures and assets/sites of customers, there are differences.

On sales side it is to be taken into account, that a not type related good is not to be sold with the original product description, since this implies a certain quality. In the sales and dispatch papers it must be mentioned, that this product is a not type related good (for this customer). In general, the code 'not type related' is set behind the original description.

Goods, which are only for certain customers not type related, remain in the original material master and are marked with a quality code. Hereby goods, which are for all customers not type related are posted in a special material master by the quality assurance. This material can, if required, be evaluated with a lower material price.

Among the batch of a material, the granulate can be presented in different packaging types. A possibility to divide a batch in the sense of the packaging, is the bin location.

6.4.10   Dangerous goods

The status sales in the material master is applied for the identification dangerous goods in the view of sales 1.

If a new material is generated, the responsible employee determines the life cycle phase (see below) and sets the sales status to 'to check'. A workflow to the dangerous good representative is generated with the instruction, that the material is to be checked. If it is a dangerous good, the dangerous good representative must switch the status sales to 'dangerous good'. If it is no dangerous good, he/she changes the status to 'no dangerous good'. The sales status 'dangerous good' controls, that with the creation of inquire, quotations, orders, delivery schedules, contracts, returns and deliveries free of charge, the warning 'material XY has the status 'dangerous good' is indicated. A delivery block indicator 'dangerous good' is automatically set for these documents, so that no delivery can be created. After the saving of a document with dangerous good, the dangerous good representative achieves on the other hand the instruction to check this document. The sales status 'to check' has the same effects on the delivery blocks in the sales documents.

The dangerous material number (MARA STOFF) with which the connection to the dangerous good database is created is registered in the material master in the delivery view. The dangerous good indicators are created respectively changed in LOGISTICS → MATERIAL MANAGEMENT → WAREHOUSE MANAGEMENT → MASTER DATA → DANGEROUS GOOD. With the help of this number/indicator, the material is identified as dangerous material or good, and dangerous good-/material data are referred to. The dangerous material/good data can be specified regional related.

6.4.11   Delivery site

A product is in general manufactured in one enterprise. The delivery site can be recorded in the material master (sales 1). It is overtaken in the position by the order registration.

The delivery site, together with the dispatch conditions (of the customer master) and the freight groups, serves for the dispatch position finding. Therefore it is defined as must field in the material master.

6.4.12   Freight groups

The freight group is to be registered in the material master (sales site data). The freight group serves as arrangement of material, which is, shipment related, referred to the same requirements. The freight serves for the dispatch position finding in the order, together with the dispatch condition (of the customer master) and the delivery site (of the customer/material/info, the customer master or the material master).

6.4.13   Quantities and quantity units in the material master

The basis quantity unit in the material master is basically kg. In the sales quantity unit a unit can alternatively be registered, which shall be proposed in the sales document position.

Any number of quantities can be converted (SUPPLEMENT → QUANTITY UNIT), e.g. pallets in kg. If a minimum order amount is given, the system reacts with a warning (or if required with a E-message) with the creation of a sales or dispatch document. With the delivery unit is controlled, whether the quantity is a multiple of this unit.

6.4.14   Packaging unit

The packaging unit is to be identifiable for evaluation purpose in the material master. The information in material short text is not sufficient for that.

If the pick-quantity (=quantity to be delivered to a customer) in the delivery is registered as differing from the order quantity in the order, the group and sub-delivery tolerances control, whether the order is finished or whether a further partial delivery is necessary. The employee is warned in the delivery document about this circumstance. The control by tolerances is important for the shipment of loose goods, since here mostly differing delivery quantities occur.

Group-and sub-deliveries cannot be maintained for every material in the material master, but merely pro customer and material in the customer/material/info. The tolerances can be manually changed respectively recorded in the order.

6.4.16   Foreign trade data in the material master

The foreign trade data are represented in the material master by a few of their own. It belongs to the sales as well as to the purchasing. If, consequently, only one sales view or one purchase view is generated, an automatic creation of the foreign trade view (without data) ensues for this material. It is recommended, that the foreign trade view is maintained by thdepartment responsible. If the marketing generates a new material master, a workflow creates a mail to the customs department with the order to maintain the foreign trade view. In the first place foreign trade data refer to:

- Country of origin of the good
- Region of origin of the good
- Export/import group: Group of materials, which show, referring to its exports or imports, similar requirements
- Preference indicators in the export/import
- Deliverer declaration code and date
- Negative-statement-indicator in the export/import
- Indicator-military good

6.5   Material finding

6.5.1   Definition

The material finding represents the replacement of materials in sales documents within the sales procedure

6.5.2   Usage

The system can e.g. with the creation of a customer order substitute a material on hand for a comparable material (e.g. the same material with different packaging). During the Christmas time is e.g. a special Christmas packaging chosen.

Further more the material finding can be applied in following cases:

- Substitute of material numbers of one's own for customer specific material numbers
- Substitute of special material numbers for European articles number (EAN numbers)
- Substitute of discontinued materials for topical materials

6.5.3   Description

The material finding is controlled by generating master records, in which it is determined, which material is to be replaced by which other materials and in which period time it should happen. The replacement reason is mentioned if required.

The product selection enables the procedure of sale of products in various forms, during the order handling. A product e.g. is sold in a line of different packaging. During the order procedure, the sales employee can select the packaging corresponding to the customer wishes. The product selection can be manually or automatically performed. Depending on the systems installation following is possible:

- A line of alternative products is offered for selection to the employees entering the order

- The system automatically selects products according availability and priority

6.6   Material listing and exclusion

6.6.1   Definition

A material listing is an agreement with a customer about the reference of a group of materials. With the material exclusion certain materials are excluded from sale to a customer.

6.6.2   Usage

The sale of material to a customer can be controlled by the material listing and exclusion. If a material listing is generated for a customer, the customer can merely purchase the materials on the list. The master record material exclusion includes those materials not to be purchased by the customer.

Material listing/exclusion is meaningful for a wholesaler, who merely deliver special products to special customers. An exclusion of the export of certain products to certain countries is performed by the export function.

6.6.3   Description

Material listing/exclusion are valid only for a special period depending on the customer's material and type of business. The material listing refers to two partner roles in a sales procedure: The ordering party and the payer. The ordering party and the payer are not identically, it is first of all checked, whether a listing is already in existence for the ordering party. Is this the case, the system checks vs. the listing and performs no further inspection. If no listing for the ordering party is existent, but one for the payer, this is checked vs. that listing. If neither for the ordering party nor for the payer a listing exists, the customer can order any material.

6.7   Collections

6.7.1   Definition

Collections are generated, if certain materials are often ordered and delivered in certain quantities.

6.7.2   Usage

The order registration can be performed more efficiently by collections. Collections e.g. can be taken over in an order document or chosen from a selection list. Herewith the current material master record is always considered. Positions, proposed of a collection, can be changed in the order anyway.

A collection e.g. which includes material often ordered can be allocated to this customer. That collection is then automatically proposed with the order registration.

If the customer usually orders the same material, it can be referred to a former order, respectively worked with a model order.

The collection maintenance characterizes an additional maintenance expenditure.

6.7.3   Description

With the creation of collections different materials can be registered with quantity proposals.

6.8.   Customer master data

Customer specific data are the most important both for the sales and the accounting. Because of that, both business departments have access to the same customer master record. The following customer master concept concentrates on the aspects of the sales view in the customer record.

6.8.1   Definition

A customer is a business partner, who places an order, to deliver or produce certain goods or services generally against payment.

All data necessary about this customer for the procedure of business transactions are stored in the customer master.

It has to be distinguished between general data, company code and sales specific data. The company code specific data depend on the company code organization. They are separately defined for the single sales fields. The general data are independent from the company code and the sales fields. They are related to a customer in all company codes and sales fields.

The general data include e.g. the customer address and the communication data.

The company code data comprise the bank and payment data merely relevant for the accounting.

The sales specific data concern e.g. the pricing, delivery and messages. These data only affect a certain sales field and are therefore sales structure dependent.

6.8.2   ## Usage

The procedure of transaction can be enormously reduced by consequent maintenance of the master data, since those are automatically taken over into the transaction. A customer is of great importance for the departments accounting and sales, only if the customer sales data are registered, the sales transactions to the customer can be recorded. The transaction can only be invoiced, when the payer is maintained out of accounting view.

For accounting data and the sales specific data are stored in the master data in order to avoid data redundancies.

General data

General data of the customer master are independent from the company code and the sales field. These are data, which are valid all over the company. Such data are e.g., the address, (communication), control, marketing, payment, unloading points, foreign trade and the corresponding partners.

Company code data

These data are applied by only one company code. Hereby it is referred e.g. to information about trade credit insurance or the account management, that is data being of great importance for the accounting.

Sales data

Hereby, data for the sales are concerned, e.g. for the pricing or dispatch conditions. They are applied only by the sales field, consequently are depending on the sales organizations, sales path and sectors.

6.8.3   **Description**

Creation of customer master data

In general the customer master data are recorded by the marketing. The initiation therefore is usually given by the field service employee. It is also possible, that the customer directly announces himself to the office work. With the introduction of the system, the data are taken over off the present systems (see here for section 6.10 'master data assumption foreign system').

Information is collected about every new customer (credit worthiness) and the field service checks also the production site, if required, in order to evaluate the customer capacity. The workflow is registered on part of the marketing, a mail (workflow) is passed to the finances, which supplement the accounting relevant data. In general, the prime orders are transacted merely by advanced payment. In the meantime the credit worthiness is checked. If the accounting related data have been registered by the FI, an additional mail is sent to the sales. With that the marketing is informed, about the complete availability of the customer data.

Partner roles/-relationships/-finding

At minimum one business partner always participates with the business transaction procedure. Natural or juristic persons are hereby concerned. A business partner can appear in several roles:

- Ordering party
- Goods receiver
- Invoice receiver
- Payer

The case, that the company placing the order is at the same time also goods receiver, payer and invoice receiver is the most common one.

155

The function 'ordering party' in the SAP-system includes all these roles.

But it is also common that offices place orders though the invoices herefore are sent to the head office and are paid by it. It is popular with trade business, that the trader appears as ordering party and invoice receiver, whereas the good is directly sent to his customer (good receiver). In these cases, the partner roles are distributed to different companies. An equivalent amount of master records is necessary. The connection between the single persons is produced in the customer master record of the ordering party.

Further partner roles relevant for the sales are:

- Contact partner: He is generated in the special customer master record of the business partner. No separate master record is created.

- Forwarding agent: The dispatch practice shows, that the goods receiver in general are always supplied by the same forwarding agent. The forwarding agent concerned can on view of the partner roles, be allocated to the customer master of goods receiver. The forwarding agent is then taken over in the sales documents. The partner role forwarding agent is authorized for the goods receiver in the partner finding. A varying forwarder can be stated per order position. Although the forwarding agent appears as vendor, he is allocated in the sales in the customer master record to the customer being shipped.

If the transport procedure is taken over by a service render, this information is of no interest for the company, for the forwarding agent is not to be maintained.

6.8.4   Varying invoice receiver

If the address, to which the invoice should be sent, is only varying a little from the address of the ordering party and if that address should appear in the invoice, a special invoice receiver must be therefore defined.

6.8.5   Account groups

The account group has the effect, that for the different partner roles of the customer only the screens and fields necessary are presented and ready to enter. Further more, the number ranges are determined by the account groups.

6.8.6   **Number ranges/ number assignment**

For every customer master record a definite number is placed.

6.8.7   **Unloading points**

Unloading points are recorded in the general data of the customer master data. If only the slightest differences to a customer are evident with the unloading (e.g. loading ramp, warehouse), not a varying goods receiver is to be generated, but different unloading points, per goods receiver can be placed. If several unloading points exist to a goods receiver the person in charge ordered to select one point with the order registration.

6.8.8   **Groups**

In the control view of the general data a group key can be re leased. Group evaluations are possible, if a match code has been built by this group key.

6.8.9   **Customer hierarchy**

Customer hierarchies enable the creation of flexible hierarchies for the application of customer structures. If e.g. multiphase purchasing associations or retail industry associations belong to the clientele, hierarchies can be built in order to present the organization structure of these customers.

They are used during order and invoice procedure for the pricing inclusively rebate determination. If a customer is dealt with in an order, who is allocated to a customer hierarchy, the relevant hierarchy path is automatically determined.

The flexibility of the customer hierarchy facilitates the maintenance of customer master data. If a new customer is allocated to an already existing hierarchy, all price agreements valid for the hierarchy node to which the customer is allocated are automatically transferred to the customer.

| 6.8.10 | **Customer classes** |

The customer class is applied for the classification of customers with ABC criteria. It appears in the views general data and marketing.

| 6.8.11 | **Dispatch conditions** |

Since it generally cannot be determined on time of the custom master data maintenance, how a customer is to be supplied, the dispatch condition is normally not maintained in the customer master. Further more, it is ensued during the order registration by the incompleteness protocol, that the dispatch condition is mentioned.

If the same dispatch condition should always be referred to a certain customer, the value can be registered in the customer master data. The value is transferred to the order and used within the dispatch position finding (commercial data of data in the details dispatch). The automatic transfer of the dispatch condition from the customer master to the order is therefore a potential failure cause, if this value comes involuntarily out of the customer master and the marketing employee has not controlled this value. Therefore it is recommended, to not maintain the dispatch condition in the customer master record.

| 6.8.12 | **Payment condition** |

The payment conditions are registered on the invoice screen of the sales field.

| 6.9 | **Customers/material/info** |

| 6.9.1 | **Definition** |

In the customer material info masters those data of a material are stored, which are defined customer individually. Essentially, these are customer individual material numbers, customer-individual material terms as well as customer individual data for delivery and delivery tolerances.

| 6.9.2 | **Customer material number / customer material name** |

With the order registration, positions can be registered by description of the customer individual material number. In SCREEN

→ SURVEY → ORDERING PARTY the material number of the customer information record master is registered. The special material number and name are as a result represented with the material name of the customer.

Less customers order products with a customer material name respectively number. They use in general the material names of the deliverer.

6.9.3   Over- and underdelivery tolerance

The ordered amount cannot be delivered exactly referring loose goods. Only after the weighing the exact quantity is fixed. For the termination of an order with minimal delivery discrepancies, the tolerances must be registered in the customer-info-master per customer and sellable material. The tolerances normally lie with ±10% referring lose goods and ±5% referring packed goods. The tolerances for loose goods arise unplanned with the loading. Over-and under tolerances for packed goods emerge by avoiding residual stocks respectively by fully using the transport space.

The tolerances stated in the customer-material-info are transferred with the order registration into the document and can there be headed, if required. The tolerances are copied into the delivery document. There, they cannot be changed any more.

6.9.4   Dispatch data

Is a delivery plant recorded in the customers-material-info, this is transferred in the customer order with the highest priority, if the material is also generated for this plant.

In the customer-info master delivery priorities and minimum delivery quantities can be notified above all.

6.10   Master data transfer out of foreign systems

In the scope of an implementation of the R/3 system in most of the companies the most different previous systems must be replaced by the R/3 system. With that, the necessity arises, to transfer the entire 'old data' of the company e.g. materials, customer, vendors and so forth to the new system in order to maintain the operative management of the enterprise.

These data can be transferred to the new system in different ways. They can be e.g. usually, with master record for master record, be entered into the R/3 system. Though, this is a very laborious way. Remedial measures can be taken: The SAP AG offers with the company own tool CATT-Computer Aided Test Tool- a universally available tool, which suits perfectly for an efficient and comprehensive support and acceleration of the development respectively modification of the R/3 system, the implementation process and the subsequent maintenance and further development of R/3. The tools are for example qualified for the automating and repetition of modular built operational processes of all kinds. CATT includes beyond that all functions necessary to illustrate, start, administrate and protocol test components for single transactions as well as test processes for dynamic operational and administrative processes.

With the help of CATT, even the laborious and time intensive procedure of the manual registration of the master data in the new R/3 system can be efficiently accelerated and automated since CATT offers the user the possibility to read and proceed master records of external files as e.g. Excel-tabular. This reduces the work expenditure within the transfer of the previous data to a great extent, as it is illustrated by the experiences wit the practical usage of the tool. Though, important prerequisite is that the previous data to be transferred are already prepared in detail and reliably by a table calculation and lie before in a readable form. On this way, up to 99999 external data records can be automatically transferred to the new R/3 system with merely one single CATT test run.

In this connection it may be referred to the book 'Testing SAP R/3 systems' (Oberniedermaier/Geiss), published by Addison-Wesley in the SAP SERIES. There, the proceeding of the transfer of previous data of external systems by CATT procedures in the SAP R/3 system to be implemented is, among others, described in detail.

6.11   Automatic master data structure

With the help of generated CATT procedures, it is also possible to newly generate any master data in the R/3 system. If, referring the implementation of R/3, an integration test or alike e.g. 100 customers are necessary, merely the CATT procedure 'customer to be entered' must be 100 times gone through, corresponding the data prepared. Referring to the expenditure, which would arise by the manual entry of 100 customers, this automating by CATT means a great facilitation.

Beyond that, this type of data transfer, offers a high measure of data safety. Data safety means in this context, that with the data transfer by CATT the manual entry is simulated, that is all entry screens are gone through and all integrity inspections are proceeded, similar to a batch-input.

By this means, also master data for

- User trainings
- Test scenarios
- IDES-process scenarios

can be generated in an easy and simple manner.

7   Pricing

7.1   General definition

7.1.1   Pricing

The SD pricing is a good example of how the global requirements are met with the help of comprehensive functionality in the R/3 system. The term pricing describes the calculation of prices and costs. Herewith all relevant information as e.g. basic price of the material, special surcharges and advance payments, possible freight charge etc. is considered.

The pricing for a material takes place automatically in SD on base of the already defined prices as well as surcharges and advanced payments. The data determined by the system however can for the concrete individual case be changed or supplemented.

Prices can come from a pricing list or an agreement with a customer or be defined in dependence on the material respectively the material costs. Surcharges can be generated e.g. to the customer or to a material group.

7.1.2   Condition

The conditions are price elements used within the calculation mentioned above. These are enlisted under special circumstances and influence the prices. Standard conditions are intended in the SD. But these need not to be applied; further company specific conditions can be created. Surcharges and advance payment are included as standard in the R/3 system and can be allocated to following elements or relationships:

- Material
- Customer
- Customer/material
- Customer/product hierarchy

- Customer group/material
- Customer group/product hierarchy

7.2   Condition technique

In this section price elements are described, which belong to the condition technique in the SAP standard. These elements are defined according to the operational requirements.

Here fore following activities are to be processed in the R/3 system-

1. Definition of condition types
2. Definition of condition tables
3. Definition of access sequences
4. Definition of calculation schedules

The following two pictures show in the IMG the way in the customizing of the pricing:

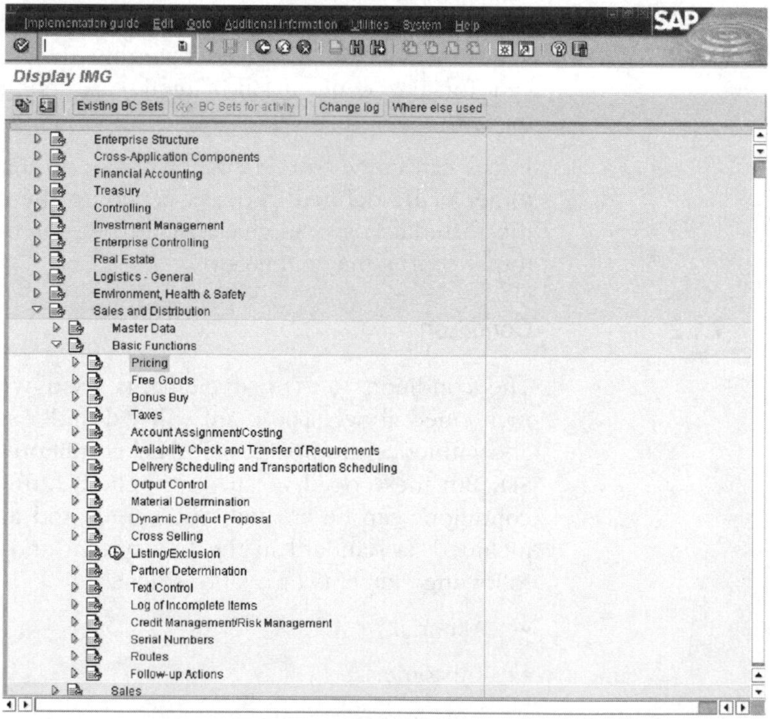

Illustration 7.2: Initial customizing in the pricing picture 1
© SAP AG

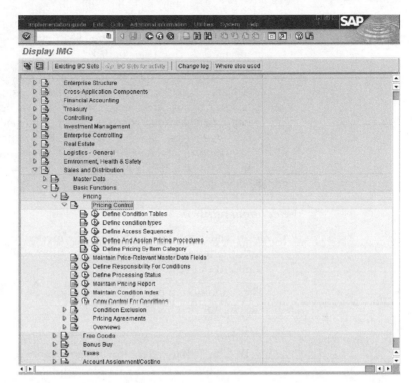

Illustration 7.3: Initial customizing in the pricing picture 2 © SAP AG

In the following terms are closer described.

7.3 Condition types

Condition types are first to be determined for all price elements occurring in the daily business and then to be performed in the system.

- Material prices
- Customer- and material advance payments
- Freight charges
- Sales taxes

The condition types can determine for a special material advance payment, that the discount is evaluated either as an absolute amount (e.g. an advance payment of 1 per sales unit) or as percentage (e.g. an advance payment of 2% for orders more than 1000 sales units).

An access sequence is given in every condition type. With that it is determined, which fields the SAP system inspects by the search of a valid condition record.

1. In the calculation scheme all condition types are summarized, which shall automatically be considered by the system within the pricing of a business transaction. It should be taken into account, that only those condition types are included in the calculation scheme that can also be manually indicated.
2. The pricing result can be manually changed in the document. The update possibilities of a condition type can be limited in this working step.

Recommendation

- If special condition types are defined, the key should start with the letter Z, since SAP keeps the name space free in the standard delivery

- The condition types included in the SAP standard delivery should not be changed

Activities

1. Test, to what extent the condition types included in the SAP standard delivery can be applied

2. A reason for the creation of a new condition type could be, that a calculation rule is necessary which is adjusted to no existing advance payment in the standard delivery. Then, a new condition type is created by copying a similar condition type. Following statements have to be made:

   - Entry of an alpha numerical key for the condition type, which can have up to 4 positions, and a text description

   - Statement of an access sequence for the condition type. (For the header conditions no access sequences need to be defined.)

3. Maintenance of the detail screen of the condition type. Herewith the possibility arises to allocate with very similar condition types, a reference condition type. Then the condition records have to be maintained only for the reference condition type.

4. In addition, there is following possibility: Upper- and under limits can be defined for the value of conditions on the level of the condition type. Herewith the amounts or

scale values are limited in the belonging condition records.

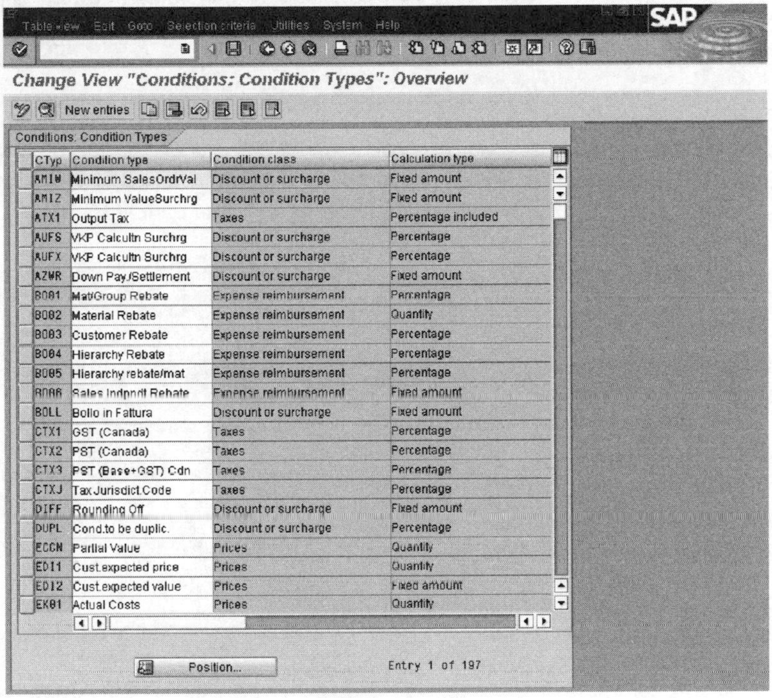

CTyp	Condition type	Condition class	Calculation type
AMIW	Minimum SalesOrdrVal	Discount or surcharge	Fixed amount
AMIZ	Minimum ValueSurchrg	Discount or surcharge	Fixed amount
ATX1	Output Tax	Taxes	Percentage included
AUFS	VKP Calcultn Surchrg	Discount or surcharge	Percentage
AUFX	VKP Calcultn Surchrg	Discount or surcharge	Percentage
AZWR	Down Pay./Settlement	Discount or surcharge	Fixed amount
B001	Mat/Group Rebate	Expense reimbursement	Percentage
B002	Material Rebate	Expense reimbursement	Quantity
B003	Customer Rebate	Expense reimbursement	Percentage
B004	Hierarchy Rebate	Expense reimbursement	Percentage
B005	Hierarchy rebate/mat	Expense reimbursement	Percentage
B006	Sales Indpndt Rebate	Expense reimbursement	Fixed amount
BOLL	Bollo in Fattura	Discount or surcharge	Fixed amount
CTX1	GST (Canada)	Taxes	Percentage
CTX2	PST (Canada)	Taxes	Percentage
CTX3	PST (Base+GST) Cdn	Taxes	Percentage
CTXJ	Tax Jurisdict.Code	Taxes	Percentage
DIFF	Rounding Off	Discount or surcharge	Fixed amount
DUPL	Cond.to be duplic.	Discount or surcharge	Percentage
ECCN	Partial Value	Prices	Quantity
ED11	Cust.expected price	Prices	Quantity
ED12	Cust.expected value	Prices	Fixed amount
EK01	Actual Costs	Prices	Quantity

Illustration 7.4: Customizing with the creation of a condition type screen 1 - © SAP AG

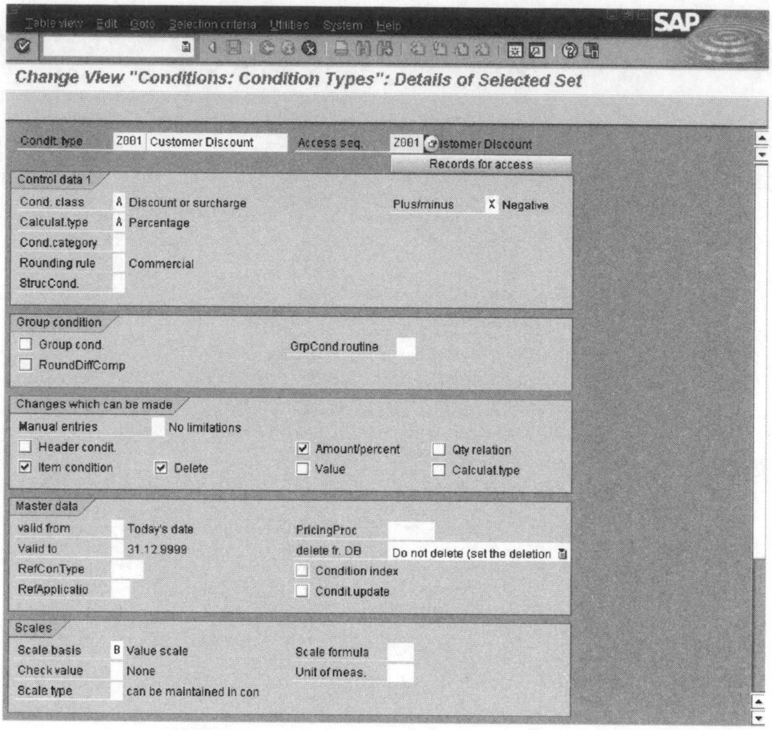

Illustration 7.5: Customizing with the creation of a condition type screen 2 - © SAP AG

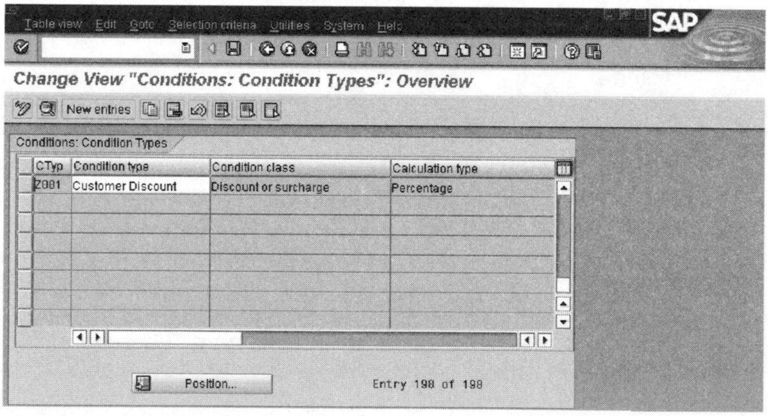

Illustration 7.6: Customizing with the creation of a condition type screen 3- © SAP AG

7.4   Condition tables

With the help of condition tables condition records can be generated for every condition type. Such a table includes following elements.

- The key: that is, a combination of fields, which mark the corresponding record

- The data part, in which conditions are filed

In a condition record these specific condition data are filed by the system, which are to be registered within the master data maintenance by the system.

An example is the table 005 customer/material with the key 'sales organization/sales path/customer/material'. That is, the customer specific material prices can be entered in dependence on the key.

Another example is the table 007 customer discount with the log 'sales organization/sales path/division/customer'. Here a certain percentage discount is given in dependence on the customer.

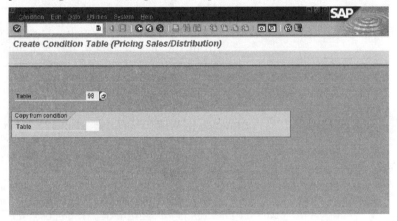

Illustration 7.7: Customizing of the condition tables - © SAP AG

In this working step the price dependencies are defined. Prices, surcharges and advance payments can be made dependent for nearly all fields in a document. These dependencies are defined by the help of the condition tables. In a condition table the combination of fields is established, for which condition records can be created.

Recommendation: The condition tables included in the SAP standard version should not be changed.

Activities to be processed:

1. Test, to what extent the condition tables included in the SAP standard delivery can be applied. The condition tables can be shown.

2. Before a new condition table is generated, it should be checked whether the existing fields of the fields' catalogue meet the requirements. If a field in the pricing is to be applied, which is not in this usage included in the field catalogue, only these fields can be accepted, which are included in the tables KOMG, KOMK OR KOMP.

3. Creation of a new condition table. Here for a similar condition table is copied and the procedure is as follows:

   • Entry of the table name; it can only be chosen names between 501 and 999. Is no statement made, the system automatically distributes subsequent numbers.

   • Statement, whether the table should be generated with or without a validity period.

   • Entry of an indication for the condition table.

   • The fields whished for the condition table are chosen from the list of fields allowed, included in the filed catalogue and there to be enlarged.

   • Creation of the condition table.

4. The condition tables are stated within their access sequences. With that the connection between condition type, access sequence and condition record is made.

7.5   Access sequences

For every condition type within the calculation scheme the access sequence is determined. An access sequence is a search strategy, with which the system looks for valid data for a certain condition table. It consists of at minimum one access. This search strategy determines in which sequence condition records are read in a condition type. Every access of the access sequence includes a condition table. The course of accesses controls the priority of the single condition records among each other. Corresponding to the fixed access sequence the system works the condition records referring a valid record. In general the access sequence is defined from the specific to the general.

Recommendation for the definition of access sequences

- If special access sequences are defined, the key should start with 'Z'

- The access sequences, included in the SAP standard delivery should not be changed

Activities for the definition of access sequences

1. Check to what extent the access sequences included in the SAP standard can be used.

2. Creation of a new access sequence by copying a similar access sequence and changing it correspondingly. In addition an alphanumerical key is entered, which can have up to four positions and a text description stated.

3. Maintenance of the accesses for the access sequence, by stating the condition tables in the sequence wished. With the sequence the priority of the accesses is determined. The therefore defined combinations of key fields can be shown for selection on grounds of the enter possibilities.

4. The generating of the accesses is from release 3.0 up not any longer necessary, it is automatically processed. Though the generating can still be processed manually by pressing button 'resource'.

Performance improvement by prestep with accesses

The accesses for a condition type can be optimized for the performance improvement, by letting the system proceed a so called prestep: It determines, whether the system shall for the time being only search for condition records with the document header data.

Is the search not successful, it is not any longer searched with this access out of the single positions. The prestep is only meaningful for accesses with header and position data. The use is the bigger, the bigger the general amount of positions in a document. (See following example).

With the pricing e.g. a customer-material-advance payment is applied. The condition records generated are based on customer data from the document header and on material data from every document position. Though, advance payment is only for 2% of the customer. So the system would search in 98% of all cases unnecessarily in every existing position. In this case the prestep improves the performance.

Several condition records are generated in the system for a rebate:

1. A customer specific rebate
2. A general rebate

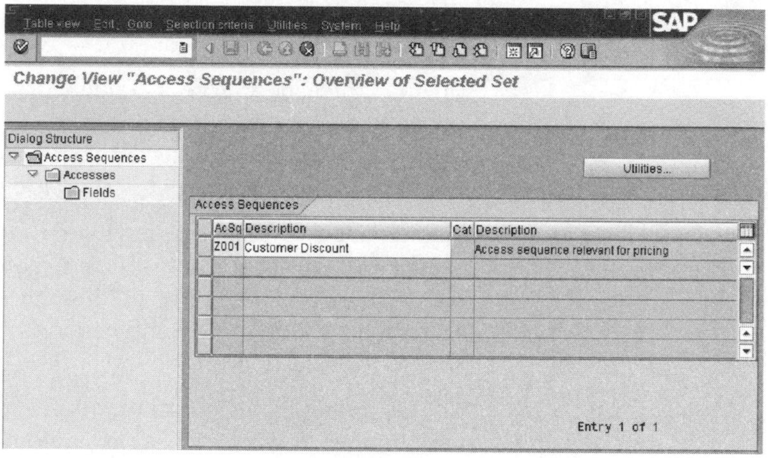

Change View "Access Sequences": Overview of Selected Set

Illustration 7.8: Customizing of an access sequence screen 1- ©
SAP AG

With the order procedure a customer can be considered for all 2 different rebate types. With help of the access sequence the system gets access to a given sequence of the condition records, until a valid record is found. If a customer specific rebate has been filed in the system, then a valid condition record would be determined with the correspondingly equipped access sequence with the very first access.

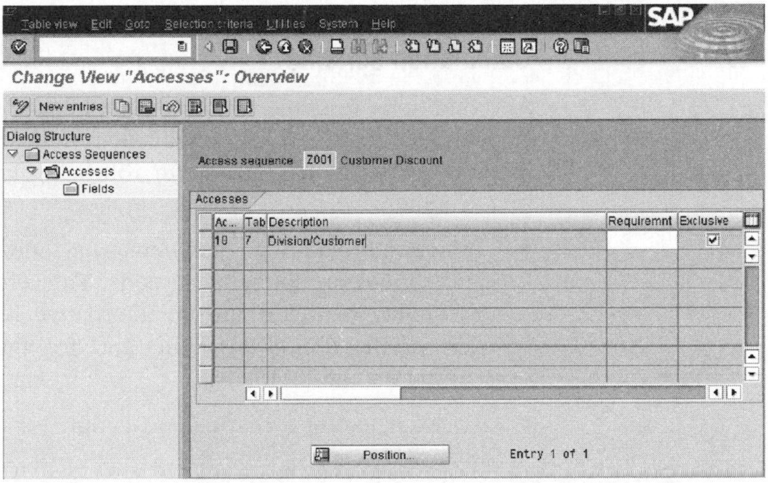

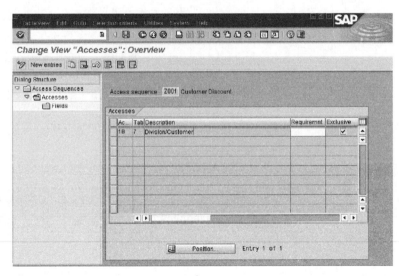

Illustration 7.9 Customizing of an access sequence screen 2-©
SAP AG

Illustration 7.10: Customizing of an access sequence screen 3 - ©
SAP AG

7.6   Calculation scheme

All conditions possible within the pricing are determined in the calculation scheme. It essentially determines the condition types in a certain sequence. Moreover following items can be defined in a scheme.

- Which subtotals e.g. between gross and net prices are built and shown on the price screens. The reference level offers a possibility to determine an alternative basis for the calculation of the condition value and for the summarization of conditions for subtotals.

- To what extent a manual procedure of the pricing is possible.

- On which basis the system evaluates/calculates percentage surcharges and advance payments.

- Which conditions must be met, so that a certain condition type is considered.

- In which sequence the condition are established in the business document.

In dependence on the customer and the order respectively invoice type different calculation schemes can be generated in the system .In the calculation scheme a condition type can be characterized as follows:

- Necessary condition

- Manually entered condition

- Statistical calculation

The calculation scheme in the TA-Order is in the further course described by examples.

7.6.1   Determination of the calculation scheme

In the calculation schemes it is determined, which condition types are to be considered in which sequence. With the pricing the system determines automatically, which calculation is valid for a business transaction and considers the condition types included one after the other. The calculation scheme determination for the pricing depend on:

- Sales organization

- Sales path

- Sector

- Document scheme, which is allocated to the customer: The customer scheme is given in the customer master record per sales field.

- Document scheme, which is allocated to the order/invoice type. The document scheme is stated for every sales document type and for every invoice type.

For the scheme determination the customer scheme and the document scheme are allocated to a calculation scheme within a sales field.

Recommendation

- In any case, special calculation schemes should be defined in which only these condition types are included, which are really used. Otherwise the system gets unnecessarily access to the condition.

- The calculation schemes included in the SAP standard version should not be changed.

Activities

1. New calculation schemes are generated by copying a similar calculation scheme. A key should be stated including up to 6 positions as well as a text description. The condition types are given for a scheme in the order of use. The lines of the calculation scheme are to be maintained.

2. In addition, the customer schemes are defined for the scheme determination.

3. The document scheme is defined for the scheme determination.

4. The scheme is allocated to the sales document and invoice types.

5. The authorized combinations are defined for the determination of the scheme.

    • Sales field

    • Customer scheme

    • Document scheme

    • Calculation scheme

•

> In the XY ltd the following calculation scheme are used for the standard orders (TA).

Document scheme	Name
RVAA01	Standard XY

Table 7.1: Calculation scheme

Following document schemes have been allocated to the corresponding order types:

Document scheme	Name
A	Standard

Table 7.2: Calculation scheme

Following customer schemes have been allocated to the corresponding order types:

Document scheme	Name
1	Standard

Table 7.3: Calculation scheme

Eventually the calculation scheme determination above results out of these factors:

VKO	VTW	SP	BS	KS	KalkS	KA
00001	01	01	A	1	RVAA01	Z003

Table 7.4: Calculation scheme RVAA01

Legend:

VKO	=	Sales organization
VTW	=	Sales path
SP	=	Sector
BS	=	Document scheme
KS	=	Customer scheme
Kalks	=	Calculation scheme
KA	=	Condition type

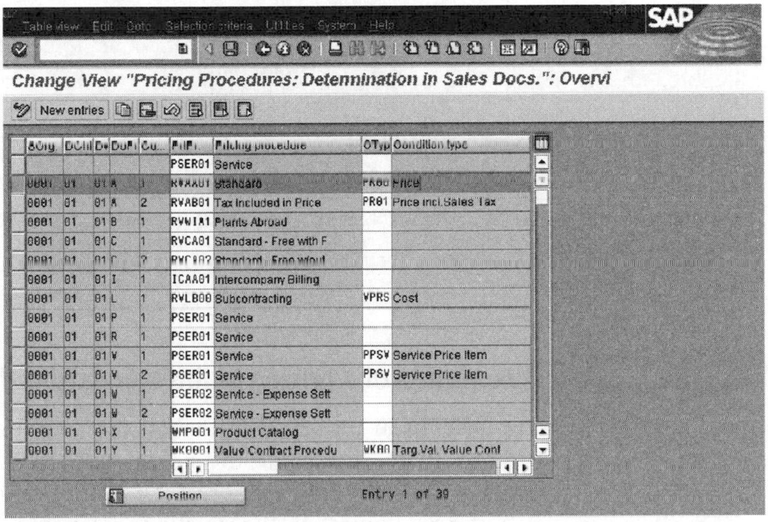

Illustration 7.11: Customizing of a calculation scheme - screen 1
© SAP AG

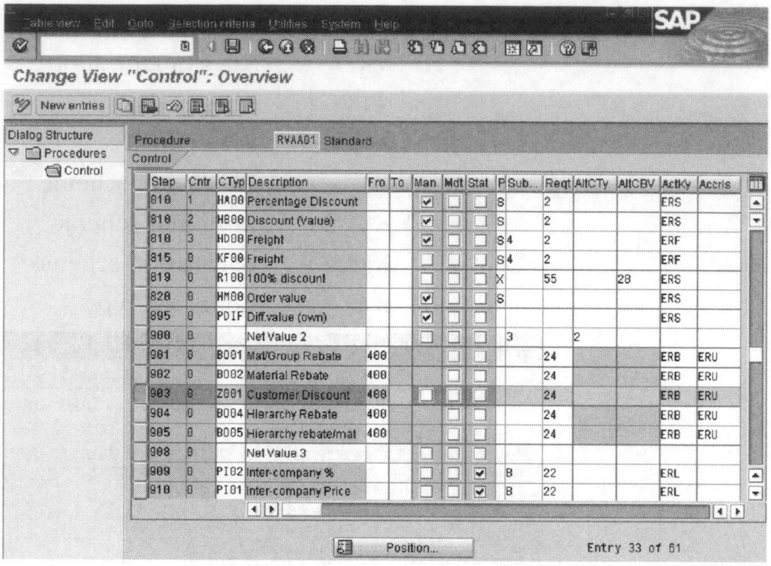

Step	Cntr	CTyp	Description	Fro	To	Man.	Mdt	Stat	P	Sub...	Reqt	AltCTy	AltCBV	ActKy	Accrls
810	1	HA00	Percentage Discount			✔			S		2			ERS	
810	2	HB00	Discount (Value)			✔			S		2			ERS	
810	3	HD00	Freight			✔			S	4	2			ERF	
815	0	KF00	Freight						S	4	2			ERF	
819	0	R100	100% discount						X		55		28	ERS	
820	0	HM00	Order value			✔			S					ERS	
895	0	PDIF	Diff.value (own)			✔								ERS	
900	0		Net Value 2							3		2			
901	0	B001	Mat/Group Rebate	400							24			ERB	ERU
902	0	B002	Material Rebate	400							24			ERB	ERU
903	0	Z001	Customer Discount	400							24			ERB	ERU
904	0	B004	Hierarchy Rebate	400							24			ERB	ERU
905	0	B005	Hierarchy rebate/mat	400							24			ERB	ERU
908	0		Net Value 3												
909	0	PI02	Inter-company %					✔		B	22			ERL	
910	0	PI01	Inter-company Price					✔		B	22			ERL	

Illustration 7.12: Customizing of a calculation scheme - screen 2: RVAAo1 © SAP AG

7.7 Description of a pricing in the standard order

The pricing in the sales is ensued in arrangement with the controlling. In the following the calculation scheme in the standard order is described by example. First the pricing scheme applied is performed and then the positions described:

7.7.1 Pricing schemes

Basic price (1)

- Freight advance payment (2)

+ Freight surcharge (3)

= *Gross price (4)*

- Rebates (5)
  immediate-, test discount, model discount, engineering industry rebate)

= *Invoice-net price (6)*

+ Value added tax (7)

- Rebate (8)

- Discount (9)

= *Order net price (10)*

- Packaging material (11)
  (incl. disposal) and dispatch packaging
  surcharge

- Freight- and transport charges (12)

- Commissions (13)
  (dealer commission, indent provision)

- Further single costs of the sales (14)
  (hedging, loss insurance, procedure costs

= *Price ex works (15)*

- variable standard production costs/landed
  price trading goods (16)

= *Contribution margin (17)*

- Silo costs (18)

- Storage costs/dispatch costs (19)

- Imputed interest/ Cost account
  depreciation (20)

- Fix standard production costs (21)

= *Contribution margin II to
  standard costs (22)*

7.7.1   Pricing scheme positions

In the following the calculation of the single price elements and (sub)totals is described. Herewith the single access sequences as well as the key and data portion of the condition records are performed in a table.

Basis price (1)

The pricing scheme starts with a basis price. This price is calculated by the marketing and pictured in the CO. From this the other values are derived. The basis price is performed by the condition type (PR00) in the system.

Access sequences	Description
1	Sales organization/sales path/ customer/material
2	Sales organization/sales path/ Price list type/currency/material
3	Sales organization/sales path/ Price list type/currency (foreign)/material
4	Sales organization/sales path/ price material
5	Sales organization/sales path/ material group

Table 7.5: Different access sequences for the determination of the basis price

Freight advance payment (2)

A freight advance payment incurs e.g. with a pickup by the customer. The refund of certain percentages for freight charges depends on the customer, the incoterms and the quantity.

Access sequence	Description
1	Sales organization/sales path/ goods receiver/incoterm

Table 7.6: Access sequence for the freight advance payment determination

The freight discount can also be guaranteed as an absolute amount.

Shipment surcharge (3)

A shipment surcharge incurs exclusively with padded goods. This short amount surcharge is raised, if the load lies under a complete load of 18 palettes. There are two palette sizes (1.250kg resp. 1.375 per palette). In the material master the palette is entered in the goods group, e.g. goods group 2.000 means the package sack. A material is exclusively allocated to one palette. Following surcharges are to be calculated:

Palette	Palette size 1.250kg (goods group 2000)	Palette size 1.375kg (goods group 2001)	Surcharge per Kg
0	<1250	<1375	1,00
1-3	1250-3750	1376-4125	0,50
4-8	3751-10000	4126-11000	0,25
9-17	10001-21250	11001-23375	0,15
18	≥21251	≥23376	0

Table 7.7: Determination of freight surcharges

The value of the freight surcharge depends on the goods group and the quantity.

Access sequence	Key	Data portion
1 Goods group	Goods group	Per kg

Table 7.8: Access sequence for the determination of freight surcharges

Gross price (4)

The gross price is the sales price referring the pricing SD. It results from the basis price minus freight discount and freight surcharge.

Discount (5)

(Immediately test discount, model discount, machine industry rebates)

The discounts depend on customer and quantity. These can be defined as percent or absolute rebate of the gross price. For the time being, exclusively rebates on position level are considered.

Access sequence	Key	Data portion
1 ordering party	Ordering party	Absolute amount per kg

Table 7.9: Access sequence for the determination of the gross
price.

Invoice net price (6)

The invoice net-price incurs of the gross price minus rebates.

Value added tax (7)

The VAT is charged on the invoice net-price. The condition type VAT calculates in dependence on the liability to pay taxes of the customer and the sales tax of the products, the corresponding tax rates.

The VAT inland for Germany incurs as follows.

Tax classification customer	Tax classification material	Tax amount
Tax free	No tax	0%
Tax free	Full tax	0%
Tax free	Half tax	0%
Tax liable	No tax	0%
Tax liable	Full tax	16%
Tax liable	Half tax	7%

Table 7.10: VAT table inland for Germany

Rebate (8)

The rebate is a special discount. This is given to a customer depending on his sale within a fixed/determined period. The details of this rebate are defined in the rebate agreement; in this agreement e.g. following data can be determined.

- Validity period

- Rebate receiver (receiver of the subsequent credit memo)

- Reference basis of the rebate (customer, customer/material, rebate group)

- Condition amount (absolute or percentage rebate)

- Accruals amount

The rebate payment always incurs at the end of a posting period. All relevant invoices (billings, credit/debit memos) are considered in the calculation. The system automatically poses accruals, so that the accounting gets a survey about the accumulated value of the rebate.

The rebate agreement is finally calculated, if a credit memo about the total value of the rebate is sent to the customer.

For the XY Ltd. 2 rebate types are provided, the amount- and value rebate. The rebate is calculated on the invoice-net price and is already revealed in the calculation scheme of the order.

Discount before tax (9)

The discount rate is calculated on the current payment conditions. Hereby the rate is determined for an immediate payment. The relate rate is listed in the calculation scheme of the order as statistical value. The reference value for the sales reduction is the invoice net-price.

Order net price (10)

The order net-price results from the invoice net-price plus the value added tax and minus rebates and discounts.

Package incl. package disposal and dispatch package surcharge (11)

The package costs for packed good result from the package price of the product, the dispatch package price, package disposal costs as well as the dispatch package surcharge.

Dispatch package surcharge

A product package combination is initiated from a standard package. The corresponding costs are considered under special direct costs of the sales respectively package and storage costs. If the customer requires another packaging than the standard packaging, a surcharge is manually to be registered.

The packaging costs depend on the site and the goods group in which the package is registered.

Access sequence	Key	Data portion
1 site/ goods group	Site/ goods group	Per kg

Table 7.11: Access sequence for the package cost determination

Freight and transport charges (12)

The freight and transport charges depend on following factors:

- Customer
- Incoterm
- Amount

Access sequence	Key	Data portion
1 customer/ incoterms	Customer/ incoterm	Absolute amount per kg

Table 7.12: Access sequence for the freight and transport charges determination

Commissions (trader commission) (13)

The commission is registered in a condition master per customer (trader). This depends on the gross goods value (gross price x amount) and refers to the order net price. The commissions are performed with rebate agreements. Provisions are built, alike the rebates.

Further single costs of the sales (14)
(Hedging, loss insurance)

These costs are considered as percentage on the gross goods value in the price calculation.

Price ex works (15)

This price results from the order net-price minus the special direct costs of the sales.

Variable standard production costs/
landed price trading goods (16)

Condition master per product. It is directly derived from the product calculation.

Contribution margin I (17)

It is the result of the price ex works minus the variable standard production costs respectively landed price trading goods.

Silo costs (18)

The silo costs arise because of the goods storage in the silo.

Dispatch- and storage costs (19)

Dispatch costs arise for the sending of goods the storage costs for the storage.

Imputed interest on UV/cost-accounting depreciation (20)

These data made available as a mean value by the CO and vary according the sales organization.

Access sequence	Key	Data portion
1 sales organization	Sales organization	(EURO)

Table 7.13: Access sequence for the imputed interest determination

Fix standard production cost (21)

Condition master per product. They are directly derived from the product calculation.

Contribution margin II to standard cost (22)

Condition master per product. They are directly derived from the product calculation.

8 Sales information system

8.1 Definition

The sales information makes compact and comprehensive information available, both to the management and the sales employees. This information on different summarization levels enables the recognition of changes in the market processes. They form the basis for premature and systematic strategic and operative decisions. The user can release the information with a less expenditure. Nevertheless, the information system has the flexibility necessary, to meet individual requirements in different sales and marketing organizations. Flexibility and reaction capability towards the market –decisive for the operational success- are optimized. An efficient sales controlling is supported by the information system and evaluations of the most important business transactions. The general question of standard business can be displayed online with the help of current information and therefore rapidly answered. In the sales information system data of business procedures in the SD are collected, summarized and evaluated. They offer various views to all information of the operative usage. Depending on the information requirement any specification level can be chosen. The data of the publishing information system can be graphically prepared. This accelerates the procedure of the information finding and simplifies the decision finding. The Sales Information System (SIS) is a flexible tool, which makes it possible to permanently control target criteria, to recognize alterations and to take qualified measures in time.

8.2 Usage/task

The sales information system, included in the SAP R/3 system as part of the Logistics Information System (LIS), serves to identify as soon as possible existing problem areas with the help of solid ratios and to analyze the reasons of their development. Basis of these ratios are operational transactions, which are performed several times a day and summarized for the use of an evaluation. The summarization of data is necessary then, when strategic decisions shall be derived from operative decisions. Hereby it is important, to recognize essential relations and trends by the help of a qualified summarization of individual information of the operative procedure.

Beside the SIS a variety of other positions and possibilities, reports, evaluations and statistics can be made available within the SAP R/3 usage on grounds of the high integration level. In this context two other possibilities within the R/3 system shall be mentioned:

On the one hand, there is, within the controlling module CO in the scope of the functionality 'Market segment and financial statement' (COPA), the possibility to call up analyses and reports for the support of an effective, result orientated sales controlling. On the other hand, information about all factors influencing the business activities of a company, can be analyzed by the usage of the SAP EIS (R/3 Executive Information System). For this purpose a SAP EIS data pool can be established company individually and provided with data of different partial information systems (Financial Information System, Personnel Information System, Logistics Information System, Cost Accounting), or but with company external data.

For the definition of future usage of CO-PA functionalities on the one hand and SIS-usage on the other hand following is to be determined:

- Sales quantity-, sales- as well as profit surveys, which deal with questionnaires of the sales controlling referring an organizational view, are to be inspected in the scope of the COPA.

- Analysis and evaluation, primarily serving for the control and pursuance of the operative sales- and dispatch business, are dealt within the SIS.

8.3   Description

The SIS serves the different decision levels of the sales as inspection and control tool and is essentially designed, for preparing information about the sales in graphical and table form.

The SIS enables nearly any view about all information of the operative use, whereby the level of information depth can be individually determined by the user.

The data analysis can follow in the SIS as a standard analysis or as a flexible one.

8.3.1   Standard analysis

The target of standard analyses is, to show up and evaluate ratios with a certain specification rate, to compare the ratios with each other and tear them up, if required, in order to display the original data of the operative usage. The standard analyses are based upon statistic files, the so-called 'Information Structures', in which important ratios are periodically updated out of the operative usage. The procedure of the standard analyses can ensue in the scope of six different SAP-specific organization elements: Customer, material, sales organization, dispatch department, sales employee and sales office. Standard analyses offer extensive possibilities of data evaluation. The following belong to them:

- Interactive selection of ratios

- Histograms

- Hit lists and ABC analyses

- Correlation

- Plan- / actual comparison

- Classification

- Flexible evaluation

- Segments

8.3.2   Flexible analyses

With flexible analyses data can be individually arranged, summarized and generated as reports. They enable the evaluation of any SAP data base structure of the sales procedure and the procurement for so-called 'adhoc-evaluations'. Following aspects are hereby considered:

- Flexible layout and text creation with the report-writer

- Definition of multi-step evaluation object hierarchies

- Data selection of the operative usage

- Definition of company specific ratios

- Operational preparation of the data as graphic

- Ratio combination of different structures

In order to create customer individual documents, first reporting objects and reports periods are stated and then either standard ratios or user ratios selected.

8.3.3   Planning functionality

For decision support the comparison with plan- and actual data is also important. Planned values for ratios are an essential factor for the construction of an efficient logistics-controlling. Therefore the information system offers, besides of actual data, also the possibility of the plan data registration and summarization. Plan values for sales, order quantities or stock levels can be planned. Those can, on the other hand be compared with actual data with in the evaluation. In this context exception conditions for the advanced warning system described above, can be calculated. The planning function is completely integrated with the sales and operation planning. All data of the actual business transactions are available. If the total potential of the integrated logistics applications in the R/3 system is used, a complete sales information system with consistent actual data is available for the standard decision process.

8.3.4   Advanced warning system

With the help of the integrated advanced warning system, exception situations can be prematurely recognized (e.g. a negative development referring the delivery period) and thereby imminent deficient developments identified and eliminated. Further more, the advanced warning system facilitates the selection of weak-points.

8.4   Requirements on the SIS

For the determination of requirements on the sales information system it is to be inspected in the particular company, which information are to be taken into consideration for the evaluation of the operative business.

Hereby it is to be considered, who requires these information and on which temporal level a cumulative information establishment is demanded.

On grounds of the different formulations of the questions and views, the requirements of the various control departments must be considered:

- Management
- Area control
- Department control

In regard of the fact, that in the scope of the available standard functionalities in R/3, a variety of ratios and evaluation functions is already available, the target is pursued in the scope of the 'fine concept' to recognize and specify possible existing contribution gaps. Requirements going beyond the standard functionality, can result from a company specific requirement and are therefore not found in SAP.

8.5   Possible requirements on the SIS

Following requirements shall vicariously be mentioned here:

* Customer monitoring / customer groups referring a quantity –as well as profit control including OPL (Operative Planning)- comparison.

* Product/product groups monitoring referring a quantity as well as profit control incl. OPL

* Evaluation of orders referring the amount and average size relation to a customer, product and/or country as well as office and field service procedure.

* Presentation and analyses of the lead times (order receipt, desired delivery data adjustment)

* Evaluation of market orientated product groups (service products, recess products…) according to different ratios

8.6   Ratios required, summarization rates, period type

On grounds of the evaluation requirements mentioned above, the necessity for the supply of ratios and measurement categories, summarization features and the determination of periodical update arises.

Ratios form in a way the data basis for the analyses-, evaluation and report function set upon. Ratios are management-interesting values, which are read out of the information structures (e.g. sales quantity, turnover).

Information qualified for the summarization of ratios (e.g. customer, material, sales organization…) is called feature. Every feature includes feature values, that is the specific information content (e.g. the customer Smith plc is a feature value to the feature customers).

Features and ratios form in general information pairs, whereby the ratios present the information value and the features the information key referring these values.

8.6.1   Features/levels

- The entire company
- The company on level of possible national companies respectively production companies
- Area levels
- Department levels
- Customer (ordering party): Customers are exclusively allocated to one sales organization
- Customer groups
- Application field: The application field describes who is responsible for the maintenance of a product and thus has the responsibility for this product.
- Market segments with order positions (level customer product): The term 'market segment' is understood as the customer specific application of a product; products, allocated to a special operation area, can find different application by single customers. With the order registration the customer should be allocated with the special positions, that is, per product, to a market segment.
- Product type (product group I): The product type describes a particular product independent of the packaging material.
- Product evaluation/positioning
- Products shall be evaluated and grouped referring different positioning on the market (Example: Standard products, project products, recess products, service products).
- Order reason with complaints (level customer-product). The order reason with returns, credit/debit memos or redelivery free of charge should serve for a higher transparency of connected costs of products and customers. With the help of this allocation a profitability evaluation of particular businesses with specific customers or products shall be performed in the future.
- Country/regional market (of the goods receiver)

- Economic region
- Integration of countries to geographical/political regions as e.g. West-Europe (WE), Eastern-Europe (EE) and overseas (OS)

8.6.2   Essential ratios

Following ratios shall be mentioned here:

- Sales quantity
- Net sales
- Gross sales
- Order quantity
- Order position quantity
- Order position value
- Return value
- Return amount
- Return positions
- Credit memo quantity
- Credit memo value
- Delivery free of charge value
- Delivery free of charge amount
- Invoice quantity
- Net invoice total
- Gross invoice total
- Customer visits number
- Date of patch production

8.6.3   Period time

A monthly update of ratios and features in cumulative form meets the requirements for the business control in many companies.

| 8.7 | **Features** |

The features determined by the mentioned requirements can be nearly displayed by the SAP standard information structures.

- *Customer* (ordering party)

- *Customer hierarchy*

- The SAP field *customer hierarchy* shall be applied for the customer specific description of structure and grouping elements.

- Customer hierarchies enable the establishment of flexible hierarchies for the presentation of customer structures. They are applied during the order and invoice processing for the pricing including the rebate calculation. If a customer allocated to a customer hierarchy is dealt with an order, the relevant hierarchy path is automatically determined.

- The flexibility of the customer hierarchy facilitates the customer master data maintenance. If a new customer is allocated to an already existing hierarchy, all price agreements valid for this hierarchy node to which the customer is allocated, are automatically transferred to the customer.

- The prices and conditions are always transferred to the delivery according to the customer and product. This is in general a mere proposal of the system

- The order registration automatically offers all sales items to the entered by, he/she can select the corresponding one. The sales item is then to be recognized in the order.

- Customer class: A-, B- and C-customers

- Product type: Product hierarchy

- Economic region: Geographical regions (WE, EE, OS)

- Country: Country of the goods receiver

- National company: Operational area of the product on the market

- Order reason with return entry referring the specific order type

- Sales-outdoor organization: Differentiation by sales offices (field service)

- Sales-indoor organization by salesmen group or possibly by the direct entry of a responsible indoor office employee with the order registration.

8.8   Analysis-, evaluation- and report function

The evaluation of the features and ratios determined are, with two exceptions, practicable by the standard analyses of SIS:

In SIS following standard analyses are provided:

- Customer analysis (based upon the organization structure S001 with the features: Sales organization, sales path, sector, ordering party and material)

- Material analysis (based upon the organization structure S004 with the features: Sales organization, sales path and material)

- Sales organization-analysis (based upon the organization structure S003 with the feature: Sales organization, sales path, division/sector, customer district, ordering party and material)

- Dispatch department-analysis (based upon the organizational structure S005 with the features: Dispatch, route, forwarding agent, receiving country)

- Sales employee-analysis (based upon the organizational structure S006 with the features: Sales organization path, sector, sales employee, ordering party and material)

- Sales office-analysis (based upon the organizational structure S002 with the features: Sales organization, sales employee group, sales office, sales path and sector)

The standard analysis can be called up with the help of different analysis views (breakdown, classification...), performance types (lists, or graphics) and analysis functions (ABC-analysis).

8.8.1   Display formats

Lists

With the help of different functions the evaluation lists displayed can be specified in relation to certain criteria.

Graphics

To gain a rapid survey, data of all list grades (referring to the specification rate) can be displayed in graphical form.

Graphical performance

- Time series graphic: With a break down (specification) referring to periods, a time series graphic can be called up which enables the display of an unlimited amount of periods.

- Portfolio graphic: The portfolio graphic gives a survey of the concentration of features in a displayed list in regard of two ratios.

- Amount/total curve: Referring one ratio, data of every list level can be presented in from of an amount curve. The graphic gives a survey, of how strong the portion of a ratio (sales) is concentrated on feature values (particular customer)

- Correlation curve: With the help of the correlation curve interrelations and connections in reference to several ratios can be recognized. For that all ratio values are standardized to an interval between 0 and 1 and the amount of the feature values selected is compared (Example: Performance of the correlation between order receipt and sales of a customer distributed among the single periods).

8.8.2   Analysis function

- ABC Analysis: With the help of the ABC-analysis a feature value can be classified in regard of the importance with *one* certain ratio. The ABC-analysis allows the setup of priorities by the following three divisions: A-important, B-less important C-relative important.

- Classification: With the classification ratio features (field contents) are distributed referring to one ratio in classes. In this way it is possible to gain a quick survey aver all features values to that ratio and to perform trends and correlations. The division of the class quantity and- limit ensues automatically, whereby the system related requirements (division in six classes) can be individually changed.

- Segments: They can divide the feature values in classes in regard of *two* ratios. The same class divisions as with the *classification* ensue system related, individual adjustments are possible. Example: Customers can be divided in classes referring the order quantity and sales → Evaluation of customers with a less turnover, but a high order quantity.

- Previous year's comparison: The previous year's comparison of *one* ratio out of a displayed breakdown- or basic list is practicable.

- Comparison of two ratios: Comparable is a ratio of the current list displayed on the screen with another ratio. Hereby, both ratio values as well as the difference (absolute and percentage) between them are displayed for every feature value of the corresponding list. These ratios are comparable, which posses the same units (currencies, quantity unit or no units). Example: With a material analysis the ratio 'Gross invoice' can be compared with the ratio 'Turnover'. The difference between both values for example indicates the quantity of discounts.

Advanced warning system

The advanced warning system is based upon the ratios of LIS and can be used for all applications of logistics, whereby a decision orientated selection and inspection of weak points within the logistics is possible. The advanced warning system enables the search for exception situations and helps to recognize prematurely threatening undesirable developments. The exception situation can be defined in form of 'exception' whereby also the conditions for the subsequent procedure can be determined. An exception consists of the statement about features respectively feature values (customer, material...) and conditions.

- Exception situations in standard analysis: Exceptions can be stated with the performance of a standard analysis. In the standard analyses the exception situations are marked in color, whereby a differentiation of the color in reference to the condition is possible. Exception situations are also displayed on higher aggregation levels and can be analyzed with help of the breakdown function (specification). Example: Exceptions on material level are already evident on the level of the sales organization.

- Exception analysis: During the exception analysis performance only these data are displayed which are related to an exception situation.

- Periodic analysis: With the periodic analysis the existent data stock is systematically searched for exception situations in a freely eligible temporal frequency. In an exception situation appears, information can follow by mail, telefax or workflow.

- Threshold value analysis: With the threshold value analysis a threshold value and a corresponding operator ($>$, $<$, $=$, etc.) is indicated for one ratio. With the help of this threshold value and the matching operator forecasts can be made, whether on grounds of present actual data a given threshold value can be obtained, in the future. The amount of the periods to be prognosticated is freely eligible and takes place on grounds of past data.

- Trend analysis: With the trend analysis the data stock can be examined with regard to a positive or negative trend referring one ratio.

9 Sales and production planning, availability check

Once the results of the medium term sales- and production planning are determined, the question of the goods availability is posed anew for every order receipt in the operative sales. Depending on this availability and various other criteria, the customer orders must then be made available.

With the availability check in the module SD the demand transfer can also take place that is the disposition is informed about the quantity required by the sales/distribution in order to be able to supply the orders entered. Due to the integration of the R/3 system the information exchange necessary between the applications Sales (SD), Materials Management (MM) and Production Planning (PP) can follow automatically. The demand is then reported in form of individual or collective requirements.

This chapter shall therefore give a brief insight for a better understanding into the sales- and production planning and illustrate the terminology within the SAP R/3, before dealing with the availability check and the demand transfer in detail. Hereby essential parts of the SD, MM and PP are cooperative.

9.1   Sales planning

9.1.1   Usage possibility

The operative business planning serves for the determination of medium term targets. Basis here are the strategic targets of the company. In the scope of the operative business planning, a flexible requirement of planned (efficiency- and consumption) quantities and values derived from these, follows for clearly defined periods (in general one year).

The targets of the operative business planning are following:

- The designed structure of the company future for a specific period. Exact requirements and targets are to be defined. The company internal and external (market-) conditions as well as the strategic business planning are to be correspondingly considered.

199

- Standards are to be created for the control of business behavior during a current billing period. In the scope of a dynamic planning, target requirements can be adjusted to the changing 'outline' condition.

- A profitability control shall be possible after the termination of the billing period with the help of plan-actual respectively target-actual comparisons.

- A requirement of a basis for the evaluation independent of accidental deviations of business efficiencies shall be created.

9.1.2 Prerequisites

- The requirement values and target levels of the planning are to be determined in cooperation and synchronization with the responsible persons. Only if the planning is accepted by them, the business procedure in the single company areas can be effectively controlled and a satisfying profitability control made.

- Starting point for the operative business planning is a sales plan, in which it is determined, which amounts are to be disposed on the market in the planning period.

- A major part of the functions of the operative business planning are covered in the planning functions of the module CO as well as in the module PP. By means of the integration of R/3 the consistent billing of values and amounts is guaranteed.

9.1.3 Procedure

- The sales plan can either be created by the financial statement or the sales information system. The planned sales quantities are transferred to the PP. In the PP the capacities and the quantity related demand on raw materials and operating supplies is determined. The planning activities respectively services required established in the production plan are transmitted to the cost centers, which have to provide these capacities in form of activity units. Further more, the cost center responsible employees must plan the incurring costs and quantities taken of other (dependent) cost centers on basis of the capacities and activities planned.

- With the planned price iteration, which concludes the cost center planning, the data are made available in form of planned cost records, in the scope of the pay scale evaluation. These are used in sequel for the evaluation of the quantity structure in the standard costing. The calculation results for these products, whose sales quantities have been at first planned in the financial settlement, are now on the one hand used for the evaluation of just these plan sales quantities (profit planning). On base of these provided (planned) cost and sales quantities revenues can be determined and from that the planned distribution margins derived.

• On the other hand the calculation results serve for the evaluation of the sold articles to standard production costs with the transmission and evaluation of the invoicing from the invoicing system in the financial and market segment settlement. Besides the standard costing, the plan cost records determined by the itcration are made available for the direct cost allocation in the overhead cost management and for the evaluation of the 'activity consumptions' in the production (preliminary costing and completion confirmation of the production order).

• There are interlockings with the financial planning, which restrictively affect the production- and cost center planning.

9.1.4   Result

Following values and quantities are now available among others:

• Planned costs and sales quantities as well as revenues
• Data for the capacity supply
• Data for the cost center planning budgeting
• Data for PP
• Pay scales respectively plan cost records
• Plan distribution margins
• Information for the investment planning

The operative business planning enables the determination of the planned amounts and supplies the data, which enable an efficient financial planning. Other business areas can also start their planning as e.g. the marketing with catalogue prices.

9.2   Production planning

9.2.1   Tasks

The market orientated production planning is given in case of an advanced planning, which doesn't depend on single customer requirements respectively orders, with independent requirements performed and which is exclusively orientated on the requirement dynamic of the market.

9.2.2   Features

The features hereby are following:

- Sales- and rough planning most important with project groups. Customer order anonymous advance planning
- No customer- or customer order reference in the planning
- Collective production with stock structure
- Processing of the customer orders exclusively out of the warehouse stock.

The plan data of the profit planning or of the sales information system are transferred as database to the sales- and production rough planning. For the products of these scenario link term planning views with following targets can be established:

- Processing of the capacity adjustments with rough planning split
- Adjustment of the cost and activity types
- Evaluation of the stock and requirement situation with long and medium term preliminary planning.

If customer requirements (sale from stock) and preliminary planning differ widely from each other, a change of the replacement planning is to be made, since an automatic adjustment and billing of the customer requirements doesn't take place with the preliminary planning.

Following process groups are part of the scenario:

- Sales- and production planning
- Forecast
- Program planning
- Long term planning (simulation)

- Major sales item planning
- Requirement planning
- Capacity planning

9.2.3   Planning strategies

Following preliminary planning strategies are assigned for this scenario (selection).

- Anonymous warehouse production
- Lot production
- Preliminary planning on assembly level
- Gross planning
- Warehouse disposition

9.3   Availability check

The sales- and production situation in a company underlies permanent changes. In the system SD an availability check is therefore processed already in the order registration in order to guarantee, that the customer needs can be fulfilled. With the shipment procedure an availability check takes place anew.

With that it is checked, whether the material is available at the desired delivery data. Beyond that, stocks can be inspected, delivery bottlenecks considered and the demand transfer processed.

9.3.1   Availability check with the order receipt

With the order receipt the sales employee has to inform her/himself about the availability of the ordered good in the warehouse. As long as this doesn't happen, the delivery date is unknown, and an acceptance towards the customer is impossible. Activities like availability, supply or transport disposition cannot be started:

On the procurement side it has to be ensued, that in case of an insufficient stock the production or purchase is informed, so that the goods can be produced or ordered in time. This information about the material necessary for the sales is transmitted to the disposition by requirements. Through the passing on of requirements the production is informed, that the goods must be produced or the purchase instructed, that order requirements exist to which orders have to be made and sent to the deliverers.

For the purpose of the availability check, the sales employee relies on the planned requirements of the sales for the actual month and the two following. The sales requirements (sales- and delivery requirements) result from all sales transactions, which transmit requirements to the material management of the production planning. These could be e.g. customer orders or deliveries, but also quotations. The sales requirements reduce an existent stock or acquisition to the material supply date, so that other outward movements elements have no longer access to the 'reserved' amount. Sales requirements are generated by the module SD, while other elements of this list are generated by the module MM respectively PP.

The planning requirements are established in the scope of a monthly planning, coordinated with the production and transformed in plan independent requirements for the production.

Through the coordinated planned figures a definite assignment of debt referring deviations is possible in the operative management, whereby the sales and production are forced to plan more efficiently and to better communicate with each other. If the sales department wants to sell more or less than it has planned, the requirement difference occupies the error planning. If the production doesn't come up or the warehouse capacity is not sufficient, the production must consider its planning strategy anew.

The processing of a passing on of requirements depends on various factors, which determine, whether and how the requirements are transmitted and billed vs. sales. Hereby the requirements transfer can be transaction related switched on or off, so that no requirement is transmitted in inquiries and quotations, whereas in orders and deliveries a passing on of requirements takes place.

9.3.2   Availability check in the shipment

On the shipment side it has to be ensued, that the forwarding agency or another transport company is informed in time, so that there is enough time for packing and loading. An availability check can be made for the deadline of the goods supply.

With the creation of the delivery an availability check of the ordered product to the picking is initiated. This check is processed by analogy to the sales documents.

The availability check in the delivery is reasonable, because of following reasons. In no availability check should be made for certain, selected materials in the order, it should be checked with the delivery creation at the latest, whether the order quantity is available, and can be delivered. But also, if the availability check has been determined with the registration of an order and an order quantity could have been accepted, it is more than reasonable to check the availability anew with the delivery creation. Since the availability check considers planned acquisitions and retirements, it is ensued, that unforeseen changes of the availability situation e.g. on grounds of lost outputs are taken into consideration.

With the delivery creation referring to an order an existent order stock is reduced in relation to the delivery amount. At the same time a delivery requirement is built, which is transmitted to the disposition. Hereby it is possible, to choose whether the delivery requirement is stated as individual or summarized requirement. Summarized requirements can be displayed for all deliveries of a day or week. The delivery requirements are considered as retirements in the availability check.

9.3.3   Types of the availability check

Two types of the availability check.

- Check on basis of the ATP amounts

- Check vs. advance planning

Check on basis of ATP amount

The ATP-amount (Available To Promise) is calculated by the warehouse stock, the planned retirements (customer orders, deliveries, reservations...) and the planned acquisitions (production orders, orders, plan orders...). With this availability check type it is checked for any transaction dynamically by taking into consideration the stocks to be considered and the planned goods movements with or without replenishment lead-time. Planned independent requirements are not taken into consideration here. The replenishment lead-time is the time necessary for the order or production of a material required. For the correct calculation of the replenishment lead-time, the times necessary must be entered in the material master record. Referring material own-procured, this is the total replenishment lead-time, in case of externally procured materials these are the planned delivery time, the goods receipt procedure time or the purchase processing time.

Availability check with incorporation of the replenishment lead- time

Here the availability is only checked up to the end of the replenishment lead-time. If the material provision date for the position of the day date lies just on the outside of the replenishment lead time, then the position can be accepted in spite of lacking availability. In this case the system proceeds on the assumption, that up to the material provision date every quantity demanded can be procured and considers the good available.

The customer wants *20 pieces completely delivered* to a d*esired delivery date*. On grounds of a backward scheduling the system determines the provision date, to which still no good is related. The acquisition of *100 pieces is completely converted* with a *retirement*, so that *no stock exists* referring to the provision date determined. If the replenishment lead-time had not been considered, following situation would result: Since the customer whishes a complete delivery, the good can only be provided shortly before the customer desired delivery date, because of the additional acquisitions (50 pieces) and retirements (40 pieces), that is to that date at which the acquisition of 60 pieces has been planned. The delivery date resulting from this provision date would correspondingly lie in the future. Since the replenishment lead-time is thus considered with the availability check, the ordered 20 pieces are provided already before that date that is to the end of the replenishment lead-time. The availability check with incorporation of the replenishment lead-time leads only then to a suggestive result, if dispositions take place periodically (with individual- and day requirements daily, with week requirements weekly), so that the amounts accepted can be compared with the acquisitions. This is, at present absolutely necessary, since the delivery date of an order, which has been accepted for replenishment the previous day, lies on the current day already within the replenishment horizon and therefore leads to a shortage. This shortage can then e.g. cause a block of the delivery creation.

Availability check without incorporation of the replenishment lead- time

If the replenishment lead-time shall not be considered, then the system proceeds an *unlimited availability check*.

Check versus preliminary planning

With the check vs. preliminary planning it is checked vs. an independent requirement, which is generated market anonymously and is in general not customer specific. The independent requirement results from the production program planning and serves for the order neutral planning of sales quantities to be expected in the future.

9.3.4   Control of the availability check

The *control of the availability* check ensues from general and sales specific control features:

General control features:

*   **Strategy group**

    In the strategy group the planning strategies accepted (main strategy and further possible strategies) are summarized. Strategy groups are allocated to the disposition group in dependence of the plant.

*   **Disposition group**

    The disposition group summarizes materials from the disposition view, in order to allocate to them special control parameters for the planning as e.g. strategy groups, billing mode or planning horizon. With the help of the disposition group the strategy group is determined.

*   **Planning strategy**

    The planning strategy determines the requirements types for the preliminary planning- and customer requirements. It is a decisive control feature for the cooperation of production planning and sales.

*   **Disposition feature and position type**

    If no requirement type is found by help of the planning strategy the system tries to determine a corresponding requirement type with the help of the disposition feature and the position type of the material.

- **Requirement type**

    With the requirement type the different requirements are identified. The requirement type refers to the requirement class and its control features.

- **Requirement class**

    The requirement class includes all control features of the view of planning purposes, as e.g. the requirement planning and –billing strategy as well as the disposition relevance. Beyond that, it is here globally fixed, whether an availability check according the ATP-logic shall be proceeded for the material, and whether requirements are to be transmitted. A precision control for the sales document is possible by the scheduling type.

Sales specific control features

- **Test group**

    The test group controls, whether the system shall generate individual- or collective requirement records in the sales- and dispatch procedure. Further more, a material stock at availability check with quantity transmission can be set. The test group also enables the *switching off of the availability check.*

    The test group specifies together with the *test rule* the volume of the availability check. It is proposed, in dependence on the material type and plant in the material master record and transmitted into the sales documents.

- **Test rule**

    The scope of the availability check is controlled for the single transaction in the sales through the test rule. Further more it is there determined, whether the check shall follow with or without incorporation of the replenishment lead-time. The single test rules decide, which stocks and acquisitions/retirements are to be considered in the availability check.

- **Schedule type**

    The schedule type controls, whether an availability check and requirement transfer takes place in the sales documents. The allocations, being possible on the schedule line, depend on the allocations in the requirement class, which is derived from the requirement type of the material.

- **Delivery position type**

  Through the delivery position type it can be controlled, whether an availability check is made in deliveries.

9.3.5   Date of the availability check

With the creation of an order, the system establishes the material provision deadline proceeding from the *customer desired deliver date*. At this moment the picking, packing, labeling and loading of the good has to start. Therefore this is also the date relevant for the disposition, at which the availability is checked.

Following data are necessary for the establishment of this date:

- The route, which determines the way between the dispatch and company of the goods recipient.

- The dispatch department, from which the good is to be delivered.

- The freight group of the material master record.

- The weight group determined in the order by means of the order quantity.

9.3.6   Scheduling

On grounds of these data having been fixed before in the system, an automatic scheduling can be processed. Since this scheduling is made backward, referring the time, it is also called *backward scheduling*. If the backward scheduling leads to the fact, that the preparations for the dispatch activities should have already started in the past for meeting the customer desired delivery date, then an automatic forward scheduling is processed by starting from the day date. Details are to be taken from 'Analysis dispatch'.

9.3.7   Availability check elements

Following elements can be considered during the availability check:

- Stocks (safety-, transfer-, quality inspection and blocked stocks)
- Acquisitions and retirements
- Orders
- Order requirements
- Plan orders
- Production orders
- Reservations
- Dependent requirements
- Sales requirements
- Delivery requirements

The sales requirements (sale- and delivery requirements) result from all transactions in SD, which transmits a requirement to MM and PP. This can be e.g. customer orders or deliveries. The sales requirements reduce an existent stock or acquisition to the material provision date, so that other retirement elements have no access any more to the 'reserved' quantity.

Sales requirements are generated through the module SD, whereas other elements of this list are created through the modules MM respectively PP.

10 Data archiving in the module SD

Economics and administration gain more importance by an ever-growing complexity. Therefore enlarging information quantity arises, which has to be faced in the most optimal way. Nowadays, in the era of Internet, Customer Relation Management (CRM), Supply Chain Management (SCM) and so forth, an efficient information processing is a competition factor of great importance.

Information and a quick access to it, becomes **the** strategic competition factor and often also to the product, which serves for the differentiation from the competition. Beyond that, rapid and efficient processing of huge information amounts plays an essential role, even during the mastering of environmental problems, the improvement of product quality, the increase of work efficiency.

The data processing can often follow up the increasing requirement profile of handling information in an ever faster, more solid and efficient way. The storage, long term archiving, administration of data and the purposeful access to a great amount of data or documents cannot up to now follow this trend and only meet partially the necessities.

Here, Document Management Systems (DMS) are fundamental solution components. By now, these systems have grown and got through a long-term experience. For the IT-conception of many companies however, they still represent new grounds. The planning and implementation require basic knowledge about the mechanisms and procedures, in order to be able to efficiently perform the necessary organizational realization. A high priority is laid also upon the optimal performance of an R/3-system with minimal expenditures. This Lowest Cost of Ownership is achieved through an efficient and rapid implementation, as well as through an optimized and save productive management.

A great data volume can often lead to activity bottlenecks, which are reflected in a bad performance on part of the user and in an increased resource expense on the part of administration. Because of space and /or performance reasons, data, which are no longer necessary in the database, must be eliminated. A complete cancellation of data off the data base cannot be taken into consideration in many cases, since in individual cases it is necessary to have access to them by reading. The data must therefore be evacuated to external storage means in such a way, that they can be read again. The logical objects to be achieved are in general physically distributed over several data base tables. The corresponding data base structures, which definitely identify a logical object, perform in combination with the related archiving programs a so-called archiving object.

10.1   SAP Data archiving in the R/-module SD

Archiving of application data is an aspect, which should have already taken into consideration during the implementation of an operational application software. Since the volume of movement and master data can rapidly increase in the productive management, the question is posed, in what way the database can be permanently relieved.

In the center of SAP data archiving, the data objects defined in R/3 are placed, which have operational importance. Data objects can be generally divided in several categories:

* Master data are characterized through the fact, that they are applied long termed in the operative business, and

* Movement data, whose character is rather of a short term nature, as well as

* Database indices.

10.1.1   Master data

Master data are the operational basic data of a company. Their quality is the fundamental basis for the successful procedure of business transactions. Within the operational standard software package SAP R/3, these basic data are filed as so called master data. The master data are filed in a central position in the R/3-system and form the basis of all business transactions in the sales. Hereby, information is concerned, referring to business partners, materials, prices, surcharges and advance payments and data as e.g. routes. The validity of master data can be confined to sectors of the company organization as e.g. to the sales fields.

10.1.2	**Movement data**

Movement data are transaction related data, which are short termed and allocated to certain master data. Individual posting documents are defined as movement data. For example, movement data concerning the sales development can be allocated to the master data of a deliverer. The total sales of a deliverer are composed of data of individual business transactions, the movement data.

With the help of the aspect summarization movement, data can be reactivated or data of different aspects brought together in a summarization aspect.

10.1.3	**Database indices**

Further data exist beside the master and movement data, which are automatically and together with the R/3-database table entries deleted in an archiving course. These data are the database indices. To every database table defined in the R/3-system several database indices can be defined in the R/3-data dictionary. If a data record is generated according to this table, the data base system automatically generates the defined indices. If otherwise a data record is eliminated, all other indices are also eliminated. That is, the elimination and realization of database indices underlie exclusively the control of the database system.

10.1.4	**Decision criteria for the data archiving**

There is a line of requirements, which must be considered by the archiving. The companies of most different sectors are liable to store all business documents (invoices, orders or other documents) for a certain period, which on the other hand varies resorts specifically. Another reason for the archiving can be the usage of archived data e.g. as legal evidence in case of a complaint by the deliverer.

The requirements on the data archiving can be generally divided in following categories:

- Operational requirements
- Legal requirements
- Technical requirements

10.1.5   Operational criteria

Referring the operational view, only these data objects are to be evacuated, which are not any longer necessary in the operative business course. Therefore it must be ensured by a test logic, that merely data objects referring terminated business processes can be archived. On grounds of the application in R/3 a strong network of data objects can result in dependence on the particular business processes. Therefore it must be tested, whether the evacuation of a certain data object requires the archiving of other objects or whether other objects must also be archived.

The high integration of the R/3-system and the possible connection of data objects in component-overlapping process chains enable the subsequent processes, which on the other hand refer to the data objects to be archived.

The principle is: Only those data objects shall be evacuated, which are not any longer necessary for the operative day business.

10.1.6   Legal criteria

Legal rules are manifested for the archiving of business data, which shall guarantee the evaluation of the data archived e.g. for an auditing/tax audit.

In some countries (e.g. USA) the tax authorities claim, that data have to be stored transparently, so that the authority can evaluate data with their own EDV-means.

The legal requirements show large country-specific differences and are therefore to be checked and met regionally in order to guarantee the auditing acceptability of the information to be stored. General statements referring type and volume of the information storage cannot be made.

The requirement is made on the archiving of certain data, that these data shall any time at a subsequent date be subjected to an evaluation e.g. through the inland revenue-office. These requirements underlie the legislation of the particular country. In order to make the evaluation of archived data possible e.g. for the U.S. revenue authorities, SAP has developed the so-called *Data Retention Tool* (DART). This tool offers the corresponding functions for the establishment of transparent databases of archived data as well as for the display of these data.

10.1.7   Technical criteria

In order to also guarantee the technical view, that data archived are still to be read after a long period and that independent from the hardware used at this archiving period and the release stage of the R/3-software, the R/3-system files additional information about with which hardware the data have been written in the archive and which data structure has been applied. This happens together with the original application data.

The purpose of data archiving is among others, the restitution of the information to be stored within defined storage deadlines and access lead times. Most of all, the access lead-times play an important role in the productive process. So it should be defined before the archiving of data, which access lead-times are required in the single departments. These requirements on the access lead-times naturally have consequences on the storage medium to be selected and the type of access to the archived data.

10.2   Relevant archiving objects for archiving in SD

The data to be archived in the R/3-system are found in many relational data base tables of the R/3-data base. In the particular application of the R/3-system, operationally matching data of the most different database tables are summarized to logical units, the so-called archiving objects.

In every R/3-module are in general several archiving objects defined. A survey of the existent archiving objects is achieved e.g. through the transaction SARA. It is to be mentioned, that on grounds of the variety of existing archiving objects a variety of archiving and deletion runs is to be considered referring the data origin.

Following archiving objects are among others allocated to the SD:

- SD_COND (Condition records)
- RV_LIKP (Delivery documents)
- SD_VBAK (Sales documents)
- SD_VBKA (Sales contracts)
- SD_VBRK (Invoices)
- SD_VTTK (Transport)

SD_COND

By means of the archiving object SD_COND, condition records can be archived. At present, only condition records of the pricing can be archived.

RV_LIKP

The archiving object RV_LIKP archives the delivery documents produced in the SD-module. Deliveries should be archived before the sales documents and invoices.

SD_VBAK

On grounds of the archiving object SD_VBAK sales documents as inquiries, quotations, orders and contracts are archived.

SD_VBKA

Archiving object for the archiving of sales contracts. Sales contracts can be archived independent from the other sales documents.

SD_VBRK

With this object all types of SD invoices as invoices, credit/debit memos, reversals and so forth can be archived.

SD_VTTK

Through the archiving object SD_VTTK transports can be archived to the deliveries of the SD-module.

10.3   Archiving systems

10.3.1   Situation

In the course of time the data volume increases ever more. A part of that must be stored, because of legal and operational reasons. Efficient electronic archiving is therefore necessary. While the application functions of the archiving are integrated in R/3, SAP partner companies offer software referring the original functions of the document management, that is, storage, filing and provision of data and documents. With progressive process duration, the data amount continually increases. The matching storage, an electronic archive solution, is to be found. Otherwise, the administration expenditure increases rapidly and the responding times get worse.

10.3.2   **Document management**

Candidates for the archive are all data respectively documents (orders, invoices or complex business processes) to which an access is hardly made or not at all, but which cannot be deleted because of legal or operational grounds. These data are to be administrated by different external storage media and with qualified storage management procedures. Since beside the R/3-data also other archive data do exist, only an ingenious document-management system comes into question as an electronic archive.

Duty division with the R/3 archiving

Since the version 3.0 more and more archiving functions are integrated in the single application modules of R/3. The essential archiving-and storage management however takes place over the particular archiving system. With the transformation of documents from an R/3 module into the archiving system, these largely loose their "R/3 semantics" that is for instance the document or customer number. The identification-link between the corresponding R/3 module and the document management system is only formed by a reference key created during the archiving. The SAP archive link is the hinge between R/3 and the archiving system. This interface is in the meanwhile available in the web capable version 4.5.

Archive optimization

The improvement of the response time is one of the reasons for the transmission of data from the operational data base server in an archiving system. With the archiving system itself, the question is posed after a while about sufficient short response time. Through efficient coaching mechanisms both at the single place of work and in the archive server, the waiting periods can be optimized. This can happen on basis of statistical values (documents often applied), through floats on grounds of system features, of the particular place of work or through a particular retrieval, with which only these data can be read in an archived system, which belong to the selected document. An R/3- archiving system can also be optimized according to the particular usage purpose. So there are definitely usage constellations, with which e.g. an access to the archive is only to be made by R/3 or the archive system shall process most of all scanned documents. The infrastructure for such special cases is simpler and cheaper than those for bigger solutions, with which e.g. even not R/3 can have access to the archive or any documents shall be processed. Result: It possibly is more reasonable and inexpensive to process several special archive systems synchronically than to cover everything with merely one "Super-system".

10.3.3   Archive Link as interface to the archive server

Through the application-overlapping interface SAP Archive Link, all R/3 application modules and various sector specific modules like SAP Retail or SAP Banking/Insurance, can use the functions of the combined document-management-system. The external archiving system is recently also called Content Server by SAP. The new version 4.5 of the Archive Link interface is adjusted to the conditions of the Internet computing. As protocol it is not any longer used the Remote Procedure Call (RPC), but HTTP, so that no additional client-software has to be installed at the particular place of work besides one web-browser. Among others, operations like the filing of documents in the archive, the deletion of documents, the provision of archived documents, the winding-up of references by hyperlinks and the processing of scanned documents are determined by the Archive Link interface. For the view of the different document formats (proprietary SAP-formats or also Tiff/fax-formats of scanned copies), the providers of archiving systems offer display programs. These programs can be partially loaded via net as ActiveX-Control or as Java-Applet (e.g. with IXOS-Archive). The Archive Link functions of R/3 perform generally a basic amount, over whose extent the archiving system go beyond. Additional functions refer e.g. the integration of not R/3-data in the document server and corresponding investigation possibilities. As a matter of fact, a part of the R/3 semantics, gone loose with the archiving, is "programmed anew" by such additional functions.

10.3.4   Criteria for the evaluation of archiving software

A survey of the consulting house Coopers&Lybrand (today Price Waterhouse Coopers) of the year 1998 defines following requirement-and quality criteria:

Requirements on archiving software

- Archiving of documents by scanning and subsequent storage

- Archiving of print lists of the SAP-pool

- Archiving of outgoing documents of the SAP script or other applications, connected to SAP Archive Link

- Archiving of original documents by scanning

Quality criteria

- Price

- Time expense for the installation

- Hardware costs for the archiving system

- Process able document types

- Speed of the document retrieval

- Acquisition to he archive (only via R/3 or in general)

- Storage procedures for former archives (Hard disk drive, tape and CD)

- Administration expenditure caused by the archive

- "User friendliness"

- Fulfilling

10.4   Recommendation for the implementation R/3-archive solution

The implementation of archive solutions in the scope of data archiving in R/3 environment poses both a scope of organizational and technical questions. These start with the qualified entry of the resulting volume of data and storage capacities, cover the analysis of the database, the determination of the solution and so forth. Even the selection of qualified techniques belongs to it. All this can be processed better and quicker with the help of a consultant and requires archiving specific know-how. A qualified consultant is therefore recommended-like in many other R/3-areas.

The guidance spectrum can-depending on the special experience- reach from an analysis and initial consultancy to an implementation company and special adaptation of the system to the operational requirements.

11 R/3 Module SD in the Internet

11.1 The Internet -a brief introduction

11.1.1 Change to information society

At present, that is, at the beginning of the 21st century, a continuous change of our society to an information society takes place. The new behavior of participants and affected persons referring the digital technology has therefore exceeded a critical state. Society and companies are into a far-reaching change, the change to an information society. Worldwide communication networks (Internet), mobile data processing, multi media, electronic cash, tele co-operation and so forth are going to change the business life fundamentally.

At the IT-fairs throughout the world, one comes into contact with Electronic Commerce. It has to be realized that hardly one exhibitionist has nothing to say about the subject Internet, and if it is merely his homepage. Most of all the big names of the IT-sector as IBM, Siemens or SAP have solid application examples at hand. This trend is also intensified by the corresponding reports referring these fairs, in which the subjects like Internet and E-commerce are increasingly identified as directive developments.

All those companies will be successful in future, which have recognized the importance of the information and communication technology for innovative, competition supporting applications, and use the technology correspondingly.

The protection of the company's competition power and the connected maintenance of places of work, is of central importance for the society.

It is to be strived for the preservation and improvement of the basic competencies of companies, with the help of the information and communication technology. For, on the one hand the investment and temporal requirements for new product developments increase more and more, but on the other hand the innovation cycles become shorter. The flexible orientation to new

221

market requirements and the predominance of time related aspects (Time to market) are ever more decisive competition factors.

Referring this background, the companies tend to a great extent to the implementation of the Internet technology to exceed and operationally use it. After an initial euphoria for the Internet engagement, the companies evaluate the Internet at present from a commercial point of view. Therefore, profit controls and cost-benefit analyses including costs, requirements on companies as well as benefit potentials reach more and more in the center of interest. The increasing amount of companies using standard software demands also in the field of EC for standardized solutions, which withstand corresponding cost-/benefit analyses and have to be adjusted merely company specifically. SAP with it's Internet offensive with increased R/3 functionalities may here for serve as a showpiece.

The Internet with its' different technologies has established itself in this context as a new communication medium. Marketing and sales via Internet play an ever-increasing role in the electronic commerce. An eminent sales increase is to be foreseen.

The use of information and communication technologies offers unique perspectives, which can be perceived by companies as a competition advantage, if these technologies are applied for improved commercial/operational solutions. Hereby it appears in the outline that the Internet and it's electronic business "traffic" develops to one of the most important application fields of modern information-and communication technology, since it can within seconds process information transmission even over great distances.

The Internet gains additional importance through a strong internalization of markets, the globalization of business strategies, an increasing mobility of capital and the fact that information has become one of the strongest economic factors and therefore the Internet contains also a great alteration potential for operational procedure [Perez98/17].

11.2   Support of the Supply Chain through the Internet

The participants of the entire logistics chain, which describes the whole goods and information flow from the first component supplier to the consumer, are different external partners. The activities for the optimization of the logistics chain, that is the Supply Chain, were up to now orientated to these internal "bounds" and confined to the mere optimization of company specific transactions.

The logistics chain can now be extended with the help of Internet by integration of deliverers and customers. The Internet serves as medium for an information flow controlled by itself, which supports the coordination of the behavior of the different participants of the Supply Chain. Customer can pay goods ordered via Internet, interrogate the current state of order or inform themselves about product availability.

11.3   R/3 applications for the Internet

The Internet opens a new sales channel for providers of goods and services. Because of the worldwide availability and rapid growth of users a huge amount of possible customers can be anticipated. Product information can be posed in the Internet and released by the customer. Customer orders can also be automatically accepted. Here it is possible, to directly contact both the final customers (Consumer to Business) and business consumers (Business to Business) [Perez98/206].

11.3.1   Business to Consumer

The scenario B2C describes the relationship between companies and customers and corresponds to the Electronic Retailing. Hereby, information systems, which serve for the usage and provision of information, marketing-and sales applications as well as various service functions for the consumer, are in the foreground.

11.3.2   Business to Business

This scenario is also called Electronic Retailing. In so called shopping malls many types of consumer goods can be offered to customers in the Internet.

11.3.3   R/3-solutions

For the R/3-system the SAP AG offers here for among others the Internet Transaction Server (ITS), which intelligently connects the Web with the SAP-system. The processes offered, are neatly connected with the operational information-and processing system, since only then the actuality required and individuality are met. Even the demanded integration of external events in the internal procedure processes can be realized, consequently a transfer of employee duties to the particular final user takes place.

The electronic retailing is getting ever more important between provider and customer, within the company as well as between different companies. The SAP AG offers for these three areas Internet solutions, with which new markets can be opened, sales-and commercialization costs reduced and the relationship to customers and deliverers consolidated.

These solutions are:

- SAP Business to Business Procurement

- SAP Online Store

- SAP Employee self service applications

11.3.4   SAP Business to Business Procurement

This solution presents the possibility to systematize the procurement processes and integrate these in the total flow of goods, money and information. SAP Business-to-Business Procurement (SAP BBP) is an innovative complete solution for the purchase via Internet. Every authorized employee can independently procure goods and services directly from his place of work, because of the user friendly, web-based surface offered by BBP. It is an inexpensive solution, which can be rapidly used in practice, even if no system R/3 is available in the company.

The advantages of BBP are:

- The strategic procurement of not-production related goods and services

- A reduced procurement cycle

- Transparency and central control of the entire procurement chain

- Reduction of personnel-and stock costs

- Rapid return on investment

- More "free space" for strategic decisions in the purchase

SAP BBP allows a real time integration of the purchase-and sales side and covers all steps of the procurement process:

- Generating and processing of order requirements

- Order and reservation with or without catalogue

- Acceptance and rejection

- Registration of goods receipt and services

- Invoice receipt

- Evaluations for management and performance data

SAP BBP supports the different possibilities of the catalog integration:

- The internal catalog administration for the purchase

- The integration of catalogs of different providers through internet-brokers

- The direct access of purchasers to the Internet catalog of providers

If one of the providers applies the R/3-product catalog SAP Online Store, BBP can generate an order in the system and at the same time create a customer order at the provider. This openness for product catalogs and existing operational systems is guaranteed by the new Business Application Programming Interfaces (BAPIs), which are XML- (=Extensible Markup Language) capable.

11.3.5   SAP Online Store

The SAP Online Store is an Internet application component for the electronic retailing between companies and customers as well as between companies. The remarkable thing about the SAP Online Store is the smoothing and unique integration of data in the R/3-system. Producer, mail order houses, wholesale trade and retail industry can directly distribute their products via World Wide Web with the SAP Online Store. Beyond that, the Online Store can be applied in the B2B-area. Simultaneously the purchase of goods is possible through this Internet component. If required, the SAP Online Store can be used as a mere Internet catalog without order possibility.

What distinguishes the Online Store from other virtual product catalogs? The particularity is that from the incoming order over the availability check up to the price mentioned, all data of your Internet business are completely integrated in your central SAP-system. This leads to a reduction of administration costs, data redundancies and failure probability.

The SAP Online Store offers following functions to the virtual purchaser:

- **Views of product catalogs.** The goods presentation on Internet pages is as important as in the branches of a company. The online goods catalog can be independently designed. In the Online Store products are presented as multimedial objects with texts, pictures and sound. The integrated search function, which is based on the search machine "Alta Vista" of DIGITAL, finds within seconds some 10.000 products and is able to connect different search criteria with each other.

- **Procedure of virtual shopping baskets**. The customer can put a selected product into his virtual shopping basket and look for other products of the quotation.

- **Customer registration.** New customers can be registered at the very first purchase and receive a customer number. Consequently, future purchases in the Online Store are simplified. This feature is in particular important for the B2B, since there the R/3-system can recognize single business-partners and present them individually adjusted prices.

- **Quotation and availability check.** The system generates an exclusive quotation for every customer, performs a current availability check and states delivery dates of the good desired.

- **Payment modality.** The customer can pay on invoice, on delivery or with credit card. For the guarantee of the transactions, credit card numbers transmitted online are decoded through the procedure SSL (Secure Socket Layer) corresponding to the standard SET (Secure Electronic Transactions).

- **Order state.** The customer can any time check the delivery or the processing state of his/her order. The data of the particular customer transaction are secured through the customer name and a code.

- **International availability.** Since the product catalog supports several country versions, the SAP Online Store can be made world wide available for the customers. The product catalog is designed on a broad spectrum of languages, currencies, date-and price formats.

In consequence, the R/3-system enables the access of any user or of another company on the functionality of the software system of an enterprise, in order to receive current information of any type (e.g. product catalog, account display) or to directly perform actions (order posting) [Buck-Emden99/258].

11.3.6 SAP Employee-Self-Service Applications

The SAP ESSAs engage the employees more intensively and relieve the HR from routine duties. ESSA is a solid solution, with which the employees gain responsibility over their data without making a special R/3-training necessary. With SAP ESS the temporary user can directly maintain personnel economic data in the R/3-system through an intuitively usable web-browser. SAP ESS provides functions for the employees, which support the management, the travel management, work flows, lists of employees and office applications and therefore go beyond the traditional Human Resources management. Every employee can obtain information through an Internet surface in R/3 or perform campaigns for his/her own. That is, a part of R/3 is relevant for all employees of a company, not only for the operational-orientated ones [Buck-Emden99/259].

Examples for ESS applications are:

- Time recording
- Internal order requirements
- Events calendar
- Vacation request
- Travel expense accounting
- List of employees

11.3.7   Internet application components of SAP

Companies, extending their business processes to Internet, can profit by this young medium to a great extent. From quotations on a homepage over the internal, IT-based order processing up to the delivery, the entire business process chain can be supported solidly. The speed and quality of the business processes can be optimized by the Internet applications of R/3. The result: Lower costs and additional capacities for the supplementation of the activity range of a company.

With the help of the so called "IACs" (Internet Application Components), which are provided by the R/3-system, it is possible to optimize business processes within a company, to extend also beyond the company limits. IACs are R/3-solutions ready for action, which can be used with every internet-browser. They are delivered by SAP predefined in the standard and can be adjusted to the particular corporate identity respectively applied as model for "self" developed or further developed application components for the Internet. The user can apply the R/3 components either immediately in their delivered form or use them as model for self-applications. Since the R/3-release 4.6 about 120 Internet application components are available, among others:

General Functions
Internal cost allocation
Internal price list
Acceptance of project data
Asset information
Integrated entry/procedure control
Procedure state report
Web Reporting Browser
Business Object Browser
Project documents

Table 11.1: Internet application component for general functions

Employee Self Service (HR)
List of employees
Job offers
State of request
Time sheet
Event calendar

Table 11.2: Internet application components for Employee Self
Service

Employee Self Service
Applications (R/3 user)
Subscriber amount (R/3-user)
Subscriber amount (WWW-user)
Reversals (R/3-user)
Reversals (WWW-user)
Address dates
Data about relatives
Tax-legal information (Canada)
Information about tax-deduction(W4)
Address statement in case of emergency
Persons called in case of emergency
Bank data
Personnel data
Recruitment

Employee Self Service
Information about
Employers contribution
Declaration
Working hours sheet
Time registration
Time balance
Working hours schedule
Vacation request
Verification of business travels made
Upload of the travel cost registration
Information about the calculation of salaries
Employment-and salary verification
Training rejections

Table 11.2: IAC for ESS

Financial accounting
Information about customer accounts
Asset information
Internal cost allocation
Internal price list
Invoice verification
Display of business travels made
Upload of the travel costs registration

Table 11.3: IAC for the financial accounting

Logistics (SD/MM)
Product catalog
Availability check
Customer order entry
Customer order state
Quality certification
Commission stock state
Service notification
Quality notification
Measurement and counter recording
Requirement request
Requirement request state
Collective release of order requirements
Collective release of orders
Confirmation of project data
Kanban
Project documents
Procurement via catalog
State information
Collective release of
Entry sheet for services rendered
Document search
Import procedure
Goods receipt

Table 11.4: IAC for logistics (SD/MM)

SAP Online Store
Distribution
Actions/Campaigns
Customer order
Customer order state
Inventory
Goods movement
Goods receipt
Branch order
List of range of goods

Table 11.5: IAC for the SAP Online Store

Six applications are offered by SAP for the R/3-system, which, referring to the usage field sales, are to be found in the areas B2C or B2B:

- Product catalog
- Online Store
- Customer order recording
- Customer order state
- Availability check
- Account balance interrogation

Product catalog

The provision of product information on a web homepage, is at present one of the most popular and applied possibilities of the commercial Internet applications. The product catalog enables hereby the display of the product portfolio for either pure advertising purposes or to initiate further acquisition steps. The catalog needs no text entry and offers a search machine and allows the connection with any graphics and texts.

Customer order processing (SD)

On the field of E-commerce, there are in the meanwhile a lot of solutions for processing customer orders. If the processing of such orders does however not take place in the productive business system of the company in which all necessary customer information and orders are handled, this is unfortunately connected with a high and in addition to it, unnecessary effort. Through the double processing required of customer data, is then e.g. a huge data redundancy generated, which should actually been avoided [Perez98/216].

This problem can be solved through the handling and order procedure via Internet an grounds of the R/3 application "Customer Order Processing". The customer order processing is the most important and therefore widely spread Internet application, since through its usage the possibility exists, to fill a virtual shopping basket with different products and order its contents. The customer order processing realizes a typical sales scenario, in which the customer "gains access to the electronic information-and sales room" on the homepage of a deliverer. An electronic catalog shows the product quotation available. The product descriptions contained in this catalog are not confined to a clear text, they can be supplemented by photos, graphics and audio clips and, if required, contain additional information. In this application shopping baskets, ATP-functions and a customer individual price structure are included.

Customer order state (SD)

The customer informs her/himself about the processing conditions of his/her already placed order, by means of the Internet component "Order State". At the same time the company is released of permanent state inquiries.

The results are satisfied customers on grounds of quick answers round the clock. Unlike the inquiry by post or telephone, inquiries by web are no additional effort for the deliverer.

The state informs the customer whether e.g. a dispatch is already active.

Availability check (SD)

An important decision fundament in the sales is the availability of the material. Only when the company, especially the sales employees, has a permanent access on the current availability of a product, dates/deadlines can be accepted.

233

This IAC informs the customer about the availability and delivery date of a product. The customer is, by means of this application, able to inquire the delivery date of the demanded materials by her/his own. According to the customizing setup, which is to be selected between the two extremes static and dynamic ATP-check, an alternative delivery date is proposed.

11.4   Ideal typical procedure of a sales scenario in the Internet with SAP R/3

A user calls up the HTTP-server of the virtual shopping mall with help of his/her browser and the corresponding IP-address. After that, the HTML-document of the homepage is called up on the web server of the provider and sent by Internet to the user. By means of a menu point, the user contacts the goods catalog of the provider.

With the help of existing material master, it is possible, to file complex product catalogs in the R/3-system. After the selection of the product group in the catalog, the materials to this product group are represented. The product list already contains a short description of the product and its price. In a product detail view a closer description of the product, as well as sound and picture/video material, which is administrated centrally in the R/3-system. The product information is directly read out of the system and is therefore always topical.

If now the customer is convinced of the product, it is also possible in this setup, to deposit the product in the shopping basket by clicking the Add-button. After the initiation of the Add-function, the forth frame opens with the inquiry of the customer number or the code /online store respectively customer order processing).

After the registration, the Login-button is to be pressed, in order to announce oneself in the system. In addition, the customer can also change his/her password. Referring new customers, a registration is necessary. The system generates therefore an R/3 customer master record and allocates an R/3-specific customer number to the customer. The customer is now asked to state his/her password, which he/she can also use for a new visit.

After the successful identification, the required amount is displayed in the shopping basket. It consists of a list with the already chosen products, which shall be ordered.

On grounds of the customer specific recognition through the legitimating, it is possible on this level, to guarantee individual contract conditions, according to the user profile filed and to transfer and perform these via Internet. Further on, the delivery date for the product is listed.

Now, an order can be generated or the quotation refused. If the customer places an order, the credit limit of the customer is, if required, checked through the access on the module SD, with the price computation.

The order placed by the customer is immediately available in the SAP-system as a customer order. An order acceptance follows via screen and that with the order specific number and corresponding order data. This number corresponds to the TA-number of the R/3-system. With the help of this number, the customer can, if required, determine via Internet the present state of his/her order. The order processing is automatically initiated in R/3. The warehouse stock of goods ordered is actualized within seconds, if the ordered is on stock.

With the announcement an identification of the customer takes place, so that only the customer specific orders can be checked.

After the selection of the order, detail information of the order, like order number, order date, currency, positions, product names, number of prices, prices as well as the present order state of the product, are given notice by the state inquiry.

The availability of the material is displayed. If the material is available at the desired delivery date, the delivery date is accepted. If the product however is not available in the stated amount of e.g.500 pieces to the desired delivery date, it is displayed, how much at once and to which date the rest is deliverable. In addition to that, the date is given, when the total amount is available for the customer. Additional functions in the finance area, like the inquiry of open positions and the inspection of the individual master data, are also realizable.

11.5　Scenario B2B Procurement - Procurement via Internet

11.5.1　Call up of SAP B2B Procurement

A user of company Y has decided to buy a material X via Internet. Since he/she does up to now not know a source of supply, he/she looks in the Internet for a provider. His /her employee is going to pay for the material X, since it is necessary for the daily work.

With the help of the standard-browser, he/she calls up the main menu of SAP B2B Procurement and chooses the function "*Creation*".

Features:

- Intuitive navigation by means of SAP B2B Procurement

- Terminated process for the procurement of goods and services

- Creation and maintenance of order requirements, orders and reservations

- Acceptance and rejection of order requirements and orders

- State control of the order requirements, orders and reservations

- Goods receipt at the work place and support with the activity registration

11.5.2　Creation of order requirements

An electronic catalog of the pull down menu *catalog search* is chosen in the corresponding entry mask. Because of the user-friendly procurement solution, the user has direct access to all information from the one single screen.

Features:

- Creation of order requirements for goods and services

- User-friendly, no training necessary

- Access on help is possible at any time

- Free text entry

- Direct access to nearly all electronic catalogs

- All relevant information appear on the screen

11.5.3 Search for goods and services in electronic catalogs

The user does not know the exact name of the product. He/she enters a description of the desired product in the entry mask for the catalog search. A list is presented with all suitable products of the catalog.

Features:

- Complete electronic catalog integration

- The catalog of the foreign provider is adjusted to the "Look and Feel" of SAP B2B Procurement

- Nearly every catalog can be bound in without problems

- Open catalog interface for the electronic catalog integration

- Simple and flexible procurement of goods and services

- Selection out of catalogs with the help of solid search functions

- Clear presentation of goods and services

11.5.4 Selection of goods and services of an electronic catalog

In the product survey of the electronic catalog, pictures of products as well as all important information, as product description and prices are displayed. Product X is found rapidly. The product is chosen and deposited in the customer shopping basket. If the customer changes his/her mind, he/she can put the product back at any time or put further products in the shopping basket.

11.5.5 Creation of order requirements

After the product is chosen, the palm pilot appears together with corresponding information on the creation screen in the customer shopping basket. The customer presses the order button to create the order requirement-an easy task with the help of the Internet capable solution.

Features:

- Creation of order requirements, orders and reservations
- User friendly and flexible procurement of goods and services
- Possibility of copying predefined shopping baskets
- Direct access to every listed catalog
- Placing of orders or security for a later processing (functions park and order)

11.5.6   Acceptance of orders

The customer immediately receives an order acceptance. This assists in the subsequent following of the order state. An order number is not necessary, since the order can be placed with a name, which can be used later on, too.

11.5.7   Examination of the state of an order requirement

After a certain time the customer wants to check the state of his/her order. He/she calls up the state picture and sees a list of his/her order requirements, which still have to be authorized. In order to receive more information, he/she clicks on the order requirements of the material X in the list.

11.5.8   Call up of state information

The user can call up comprehensive information about his/her order requirement as e.g. account assignment-and header data as well as information about the state of approval. Is the order requirement already authorized? Who is responsible for the approval? A graphical performance of the approval process with the current state of the order is represented on the *approval picture.*

Features:

- Integrated entry
- Workflow controlled
- Process automation
- Transfer of approvals
- Exceptions
- Definable

- Predefined workflow scenarios
- Workflow definitions can be generated and changed
- Process control
- Transparent and topical process performance

11.5.9   Acceptance of the goods entry at the work place

From the generating of the order requirement up to the delivery of the material X only a few days pass by. The customer calls up the entry screen to accept the goods receipt. With the self-service function, he/she can process everything without any problems.

Features:

- Goods
- Reception confirmation by the customer at the work place
- Services
- Registration of the made activities by the provider/ component supplier
- Automatic transmission of the entry sheet for the services rendered for the approval to the customer

11.6   Final contemplation

The present trend results in the connection of all existing components with each other through interfaces, in order to optimize the data exchange and the communication worldwide or internally as well as to integrate the operational processes. In contrast to this is the fact that many enterprises have still contact difficulties towards the electronic commerce or believe that in their sector the time has not yet come. It has to be determined, however, that the Internet book retailer Amazon sells more books today, than the US-market leader Barnes& Noble. This makes clear that companies of the most sectors become themselves victims of the new rules of the Internet. The fundament of competition changes dramatically: Customers requirements, value-added chains, sales channels and marketing strategies gain new shape and growing importance and turn many market mechanisms, which have been taken for granted, upside down. A real profit for the companies however, is only then yield by the modern architecture, when it is put in shape for individual requirements of the company and is efficiently bound in the complete strategy. This requires three decisive know how-tracks, which are already implemented in the SAP R/3-Internet applications: Internet technology, integration of heterogeneous system environments and managerial economics must be consequently brought together. This concerns also and most of all the sales.

12   ASAP – Efficient method for the implementation of the module SD

12.1   Today's Situation

At the beginning of the new millennium many companies, among others also mid-sized companies, find themselves confronted with various challenges. The rising cost-and competition pressures, as well as the increasing globalization of the markets are only examples for the developments characterizing the actual situation. Courageous and far-sighted decisions are required, in order to master this change and stand the chances occurring due to the changes. A common information procedure often rather handicaps as to promote the setting-out to new strategic shores. Referring this background, many companies of all sizes decide for a fundamental change to the R/3-standard software. The implementation of this reliable operational software, is a great chance for the companies for the business process optimization, in the sense of a best-practice-orientation. And just the simplification of standard procedures in the company provides the company with additional resources, in order to focus its engagement on the basic competences.

The new application software for companies must hereby, as described, cover a broad spectrum of functions. Simultaneously the software must be sufficiently flexible configured, to meet all special requirements, which can vary to a great extent. The numerous experiences of practice show in this context, that the periods for the implementation of the complex operational standard software package SAP R/3 vary; nevertheless, in essential dependency on the company size. The bigger the company, the longer in general the implementation process, starting with the pre survey up to the going life.

12.2   Implementation strategies for SAP R/3

With the implementation of a standard application software, as experiences of practice show that a very long, labor intensive and difficult way is to be undergone, until the manifold possibili-

241

ties of the software and the most different needs of the company are adjusted with each other in the most optimal way. Only then the system can go life. Companies, newly confronted with the implementation of software systems, therefore require a special project management for the implementation of the software.

The right choice of the implementation strategy is of central importance for the success of the standard software project. Basically, several procedures are possible:

- Immediate implementation (Big Bang) or

- Gradually implementation.

In dependence on the size and the geographical distribution of the company, the amount of business processes to be implemented/supported and the number of employees, the qualified implementation strategy is to be determined.

12.2.1   Immediate implementation - Big Bang

The advantage of an immediate implementation is, that all existing former systems can be replaced by the standard software at one time. The result is, that on the one hand the license – and maintenance costs are reduced, and on the other hand no interfaces between the new standard software and the old systems are to be built-up and maintained. The lacking interface realization reduces time and costs of a project to a great extent and spares specific know how about interface technologies in it.

12.2.2   Gradual implementation

The implementation can be gradually proceeded with limited functionality and is therefore faster productive. The display of more complex functionalities and the corresponding configuration can follow later on, if required. This gradual implementation volunteers for example in case of a high number of users, a multi- site surrounding and very heterogeneous requirements on the software on part of product- or business field specific circumstances of the company. Moreover, "experiences" with the new software can be made. On grounds of the graduation, the interfaces must thus be realized and maintained, referring further required other application systems.

If the company decides for a gradual implementation, functional dependences mark an ideal path between the part-systems of the standard software. Referring the R/3-software with its already presented part-systems it volunteers

- To first implement the operational modules (financial accounting, and cost accounting/controlling),

- Then, the logistics modules (sales, material management, production planning – and control, project system…) and

- Finally- if required – the human resources management.

On grounds of the parallelism of quantity- and value flows realized in R/3,according to which every stock-/amount changing transaction in the system releases online-realtime a valid transaction, a foreign system for FI and cost accounting must be maintained during the prime implementation of the logistics modules.

12.3   Project management and methodology usage

With its R/3 procedure model, SAP has developed an implementation methodology for its' products. The procedure model comprises the most up-to-date knowledge that SAP has gained from successful R/3 implementations, knowledge that the model makes available in the form of recommendations. Wherever necessary, the traditional model offers comfortable access to the required tools, such as the R/3 reference model or the Implementation Guide (IMG). The procedure model therefore enables ideal scheduling of an implementation project. It also builds the foundation for estimating the time and resources needed to complete a project. From the procedure model, you have direct access to a project's documentation.

The traditional SAP procedure model describes implementation like a cascade: it describes several phases that you must execute in order, according to a strictly predetermined sequence. A milestone meeting occurs at the end of each phase. The meeting informs the members of the steering committee about the results achieved in the preceding phase. The steering committee must then decide to begin the next phase, repeat the current phase, or interrupt the project.

The traditional SAP procedure model consists of

- Organization and conceptual design

- Detailing and implementation

- Preparation for productive operation

- Productive operation

All working packages in the single phases include recommendations and project administration support.

The companies, especially those of mid-size, want keep the implementation times for new systems understandably low. Nevertheless, quality and productivity shall be on highest level. SAP has reacted in time on the ambitious challenge, by making, among others, qualified methods and tools available, which should help the customer to essentially reduce the implementation times.

In order to meet the customer demands for shorter implementation phases, the SAP has developed following devices:

- The business engineer and

- The implementation method ASAP.

Both tools are precious devices for the project implementation and for the reduction of project lead times. Its application requires elementary methodic basic knowledge of the particular project team members in the fields of:

- Project management,

- Event controlled process chains,

- Entity-Relationship-models and/or

- The business framework architect.

12.3.1   Business engineer

The business engineer serves for the customer specific configuration of the R/3-system: It consists of

- The procedure model (phases and activities of an R/3-implementation project)

- The reference model (complete description of the R/3-functions, processes and business objects)

- The implementations guide (dialogue controlled customizing tool) and

- The business workflow (tool for the automating of business processes).

12.3.2   **Accelerated SAP-ASAP**

ASAP has the target, to reduce the implementation by providing

- Experience based knowledge
- Implementation recommendations
- Check lists and
- Instructions,

for a pragmatic usage of the business engineer.

With the help of this tool the implementation times can be considerably reduced. In the ideal case, the implementation times should, referring to SAP, be realized within some six to nine months.

In the following, the ASAP-implementation methodology is described.

SAP America developed ASAP in 1996 in collaboration with several experienced R/3 consultants. The often-expressed desire of numerous US customers for a significantly abbreviated implementation period became a reality with the introduction of ASAP. In Europe, SAP has responded to this wish since 1997 with *Accelerated SAP* (ASAP). With this implementation strategy, smaller and mid-size enterprises can introduce R/3 software within six to nine months.

Target of the SAP-implementation methodology is, to achieve as fast as possible a satisfying productive system and to enable system optimizations in the subsequent projects. With the version 4.5 ASAP offers support for following types of implementation projects:

- R/3-Prime implementation
- R/3-Upgrades
- Company –Rollout
- Global ASAP for distributed business processes.

12.3.3   Principle

On basis of the Continuous Business Engineering (CBE), the A-SAP-model has been developed. CBE means in this context, that the internal requirements referring to the solution possibilities are checked by R/3. Hereby the actual quantity taken respectively the creation of a prototype is relinquished. Instead of this, a version for the business processes to be supported and the organization structures is designed. Subsequently, the company determines on base of the Best Price Practice the target concept, coming into question. This target concept is referring the ASAP-terminology also called " Business Blueprint".

The Business Blueprint forms the basis for the configuration of the R/3-Baseline-System. In this Baseline-System 100% of the desired organization structures and 80% of the desired business processes are configured. The remaining 20% of the business processes are faced in a former period.

The sequencing of the CBE-steps within the implementation of R/3 is the exceptional feature. If the Business Blueprint is completely worked out and released, the system is correspondingly adjusted in the system.

In order to cost-effectively configure R/3, ASAP includes many tools for an individual configuration of R/3. The R/3- processes, as well as the components, functions and organization factors, can be shaped referring to the company requirements and unused functions can be deactivated.

ASAP is the standard-implementation method of SAP, in which the ASAP Roadmap is included. This ASAP-Roadmap is a guide, containing the experiences of many years of R/3-implementation, being established step by step. Since the R/3 –release, ASAP is integrated in the main menu of the R/3-system.

Behind the term Team SAP an initiative of SAP hides, which supports the successful usage of the R/3-system on grounds of:

- How much will the project cost?

- How long will it take?

- How do I save the quality?

ASAP answers questions referring the R/3- implementation and supports in all phases of the project with recommendations, tools and measurements, as for example:

-
- Which tools can be used?
- How long will the project take?

In this connection ASAP represents a total solution for the efficient implementation. Hereby several components are united, which support the quick and efficient implementation of the R/3-system: These components are:

- Method
- Tools
- Service

In my opinion ASAP has developed from today's classic R/3-procedure model and integrates the current knowledge of the R/3-implementation. ASAP is therefore not a totally new procedure, but provides the collected know how of SAP-customers and partners.

ASAP demonstrates the configuration transactions and enables the potential of the R/3-system to be fully exhausted, as well as the dependence of the companies to be minimized. ASAP is therefore especially qualified for following groups:

- Employees of a company, having to discus, develop as prototype and design the company model and the Business Blueprint.
- IT-departments of big companies, which shall be more effectively and quicker adjusted to the R/3-applications.
- Small and mid-size companies, which have up to the moment hesitated to implement R/3, because of the expected extents of such projects.
- Consultants, looking for effective possibilities to offer individual configuration solutions to customers or wanting to develop solutions for a specific sector on basis of R/3.

12.3.4   Role concept

ASAP arranges all duty packages beyond the single phases of the implementation referring business fields and roles. Hereby 3 roles are to be mentioned in an implementation project:

- Project manager: He/she is responsible for the planning and procedure of the project

- Application consultant: He/she creates in cooperation with the project team the Business Blueprint, defines the R/3-system and leads the training for the users.

- Technical project leader: His/her responsibility lies in the definition of the system prerequisites, the configuration of the system landscape and the training of system administrators.

With the help of this role concept, it is ensured, that corresponding to the scopes of duty, problems are solved purposefully and free from redundancies.

12.3.5   Phase concept

The ASAP-method is based on 5 phases:

- Project preparation

- Business Blueprint

- Realization

- Production preparation

- Go-Live and support

Any of these phases is allocated to different working packages, which have to be proceeded in a fixed order in the course of the phase. Hereby, concrete devices as example documents or – presentations, as well as concrete, recommended procedures are offered to the user. The working packages with all their single activities are stated in the appendix.

The ASAP-implementation method differs from others most of all because of the "Timebox-approach". The central element of an ASAP-project is its lead-time - the project size must adapt to this time limitation. That is for the total project as well as for every single project step. Therefore ASAP requires ever again quick and partially hard decisions.

12.4   ASAP-tools and accelerators

With ASAP different tools and so-called accelerators are delivered. The ASAP-tools are individual applications, which serve for the support of the performance of single working packages and or phase activities. These tools are

- Different PC-tools

- Implementation assistant

12.4.1 Different PC-tools

For the detailed procedure of the project plan e.g. MS-Project and Excel are applied. Winword serves for documentation; Powerpoint for visualization of procedures.

12.4.2 Implementation assistant

The implementation assistant is a collection of tools, which can be applied independently from the R/3-system. This tool serves as navigation tool, which accompanies the user during 5 phases of the R/3-implementation. It is PC-supported and has been developed for the SAP-procedure.

The implementation assistant is among others built up as follows:

- Roadmap
- Implementation accelerator
- Project plan
- Question and answer data base
- Issue-data base
- Business process master list
- Knowledge corner

Question and answer database

The Q&A-database is a repository of all questions and the corresponding answers of the company. These questions and answers are required for the definition of the business requirements and for the development of the business solutions in regard to the R/3-reference model and the R/3-system. Business processes, technical questions, organizational questions and these to the configuration (and certainly its answers), serving as source for the establishment of the blueprint, are considered of.

Issue database

An issue is an unforeseen activity, project or business situation, which affects business and project targets and delays the project time plan. An issue can result in changes of the project size, the budget, time plan and resources.

The fundamental duty of the project leader is to solve problems and clarify them during the project, who therefore builds the basis for a successful implementation. The main attention should be directed to the clarification of issues. Through the Issue-data base the project team can register, prosecute and document open items. The database contains following data for every registered, open item:

- Priority
- Project phase
- State
- Responsible person
- Clarification date required
- Actual clarification date

Project plan

The project plan includes 3 components: A budget plan, a resources plan and a working plan.

Budget plan

The budget plan contains the projected costs per month in comparison with the actual costs and calculates the deviations.

Resources plan

It includes the resources, which have been allocated to the R/3-implementation. It shows the planned and actual amount of working days per month as well as the deviation between them.

Working plan

The working plan includes a detailed record of all phases, working packages, activities and tasks of the Roadmap on one time scale. This information is organized as planning tool for the project administration, which can be regarded e.g. in MS-project.

Business-process-master list

This tool administrates the configuration, tests and the creation of user documents. The business-process-master list is connected with the prepared business process procedures, which serve as model for the initial definition, development of working procedures and business-/test transactions. This business process procedure provides the fundament for specified user documentations of R/3-transactions.

Knowledge corner

It contains detailed description of R/3-processes and functions, accompanied by adjustment auxiliaries for special application areas. These are very useful during the establishment of requirements and during the configuration. The knowledge corner contains tips and tricks of consultants as well as information about technical tools.

Targets of the knowledge corner are:

- The provision of detailed information and alternatives, in order to comprehend and configure the specific functionality within the R/-system.

- The provision of detailed information about further additional functions, as e.g. support-services.

- The provision of information referring to implementation activities, as e.g. data converting, approval administration and schedule development.

Roadmap

The roadmap is a project plan describing in detail in 5 phases, which working package, why and how is to be proceeded. In the following, first the ASAP-roadmap and then every single implementation phases are described.

12.5   Implementation phases of the ASAP-roadmap

In the roadmap all activities of an implementation are described. Hereby it is ensured, that none of these is omitted and that the project management precisely plans e.g. the training for users. Further more, the roadmap includes also the total technical area for the support of the technical management and deals with things like interfaces and access approvals, and this earlier as it is usual with the most ordinary implementations. The ASAP roadmap is the successor of the SAP- procedure model on basis of R/3 that was used up to release 3.1 in R/3 implementation projects.

As the ASAP Roadmap shows, this methodology includes five phases:

- Project preparation

- Business Blueprint

- Implementation

- Preparation for productive operation

- Go-Live and support

Each phase uses assigned work packages that ASAP then processes in a specific order during each phase. Each offers the user concrete assistance, including sample documentation and presentations, along with recommended procedures. An appendix lists all the work packages and the individual activities assigned to each.

The use of a timebox differentiates ASAP implementation methodology from all other approaches. A specific running time defines the central element of an ASAP project: the project scope must conform to this period. The conformity applies to the overall project and to its individual steps. ASAP therefore always demands more rapid and harder decisions.

The following briefly describes the five phases of the SAP Roadmap.

12.5.1   Phase 1: Project Preparation

This phase of the ASAP implementation methodology plans and prepares the R/3 project. Here, you consider the following points:

- Define the goals and objectives of the project

- Clarify the scope of the implementation

- Set the implementation strategy

- Set the schedule for the overall project

- Set up the organization and responsible bodies for the project

- Assign resources

This phase must process the following work packages:

- Rough planning of the project

- Project procedures

- Kickoff of the project

- Planning technical requirements

- Quality assurance and project preparation phase

One of the most important procedures is the question of the creation of project documentation, most of all the documentation to the R/3-planned concept. A qualified project management is absolutely necessary for information about project decisions, the clarification of issues or configuration changes, required to a former date. The following types of documentation should in this phase be defined and maintained during the entire project:

- Project activities lto be performed

- Project work papers

- Business processes to be implemented

- Statements to the R/3-design for business specific expansions

- User documentations

12.5.2   Phase 2: Business Blueprint

This phase aims at creation of a document that contains detailed documentation of the results determined in the requirements workshops during the first phase. The document is called the Business Blueprint. The Business Blueprint therefore documents what business processes the enterprise requires. You use the document to gain an exact overview of how the enterprise will map its business processes in the R/3 System. The Business Blueprint serves as leading target concept, also called special concept, and includes all implementation details in one single document. This document is a detail referred summarization and documentation of the business requirements and forms the basis for organization, configuration and, if necessary, development activities. The Business Blueprint ensures, that every single employee exactly comprehends the detail extent of the project in regard to business processes, company structures, system surrounding, project team training and project standards. Issues to changes in the implementation volume, effects on the budget and resource planning must be processed by means of Business Blueprint.

Here, you must process the following work packages:

- Project management

- Train the project teams for the second phase

- Define the system environment

- Set the organizational structure

- Set business process definitions
- Perform quality assurance

12.5.3   Phase 3: Implementation

The primary goal of this phase involves production of a completely configured, tested, and accepted R/3 System. This phase implements the customer requirements documented in detail in the Business Blueprint. You use specific tools to consolidate the information required for implementation. For complete implementation, you must also create programs for data transfer, interfaces, and reporting. This phase executes a closing integration test that presupposes the planning, implementation, and testing of all interfaced and system enhancements. Within the R/3 System, the Business Engineer documents and transports the configuration.

The work packages in the phase include:

- Project management
- Train the project team for the third phase
- Create and test Basis configuration (baseline)
- System administration
- Create and test detailed configuration
- Develop enhancements, reports, and forms
- Develop and authorization concept
- Set up archiving
- Execute an integration test
- Plan and implement user training
- Quality assurance

12.5.4   Phase 4: Preparation for Productive Operation

In the fourth phase, you create the preconditions for productive operation. At the end of this phase, the R/3 System should be able to process all the previously defined business procedures according to requirements.

This phase includes the following work packages:

- Project management

- User training

- System management

- Planning for the cut-over and support

- Cut-over

- Setup of the help desk

- Quality assurance

This last preparation phase before productive operation clarifies all the remaining open issues and questions. After successful completion of this phase, the production system of the R/3 System can execute business processes.

12.5.5   Phase 5: Go-Live and Support

The goals of the fifth phase include support of the operative R/3 System and optimization of system throughput. Here, you change over from a pre production environment to a production environment. Experience has shown that users have many questions during this phase. Accordingly, it is imperative to set up solid support systems for the R/3 users. Support should remain available over the long term, not simply in the first few critical days of productive operation.

The phases includes the following work packages:

- Production support

- Activities after the start of production

This phase closes an R/3 implementation project. Ideally, permanent optimization and further development of the R/3 System continue after this phase.

12.5.6   Summary

The ASAP- roadmap covers consequently the different aspects and phases of an implementation. It provides a repeatable standard procedure for the introduction of the R/3-system, which comprises the project management, the definition of the business processes and the technical as well as test-and trainings aspects.

In any case possible, the ASAP provides examples, check lists or edit formats as models.

These examples, check lists and edit formats shall be basis for the general project procedures. These models are also called "accelerators", which are used for any kind of implementation. Various accelerators are include e.g. in the so-called "Knowledge Corner". They form a collection of descriptions, instructions, models and examples for themes referring the implementation of R/3. Some are short information texts about a special subject, others are longer ones. About 350 accelerators do exist for ASAP, which can be called up in an alphabetical list.

The alphabetical list of these accelerators is added in the appendix.

12.6   Prerequisites for short introduction periods

- Clearly marked, stable project content

- Business Process Reengineering only based upon the reference model

- Implementation based upon the R/3-standard

- Solid support within the company

Target of this chapter is the development of a supporting questionnaire for the application of ASAP within the implementation of R/3. By this means, a practicable guidance should be given both the consultants and the users, which should simplify the ASAP- procedure method

12.7   Advantages of ASAP

With ASAP and R/3 costs and time can be spared, as well as the quality controlled, without having to make compromises on implementation targets. The advantages of ASAP are:

- Shortened implementation times and quicker return on investment through a structured planning and predefinition

- A clear understanding of the huge spectrum of the R/3-functionalities

- Process optimization through solid scenarios, processes and value added chains. These make the software possibilities plain and offer practical help for the definition of the R/3-system

- Qualitative implementation on grounds of instructions to be met for the procedure methods of the implementation

- Optimization of the business processes by the usage of SAP –Workflow

- Continual, dynamic adjustment and optimization of the R/3-applications

12.8 Question techniques ASAP-for the efficient SD-Implementation

Subsequently a question catalog is presented, which should serve for simplification of an introduction of the module SD referring the ASAP-method respectively for arrangement of certain solution approaches.

The questions stated there, have its' origin in following areas:

- General facts about the company
- Questions referring the sales logistics

12.8.1 General questions about a company

- Determination of the sales organization
- Definition of the sales strategy
- Description of the customer master
- Division of the customer master in certain groups
- Listing of products and services, with which the company gains its turnover
- Definition of the business areas
- Allocation of the products and services to the corresponding business areas
- Creation of a customer list

12.8.2 Questions about Sales Logistics

Questions referring the company structure

- Can an order consist of a mixture of products?
- Who registers these orders?
- Are only special kinds of orders settled?
- If yes, which?
- If several alternatives are available, referring which criteria a production site is chosen?

- The transport is organized out of which areas respectively sites?

- How are the customer master data organized?

- Who is responsible for the maintenance of customer master data?

- Which sales organizations are there?

- Which sales paths are available?

- Which sectors are there?

- How are the sales areas defined?

- Which methods of pricing are applied in the particular company?

- How is the department for the order procedure structured?

- Does a centralized or a decentralized order entry take place?

- How is the invoice procedure organized?

- Are commission agreements made in the company?

- Are sales document generated?

- Are rebate agreements determined?

- How is the customer data maintenance organized?

- How many company codes are in the enterprise?

- Are products and services of every company code offered and is a turnover gained?

- Can the entire staff register orders, or are there single specialists?

- In case of several possible production sites, is the order allocation decided upon the customer site, the production site or the combination of both?

In the following, the sales logistics is divided in the single processes:

- Customer order procedure

- Delivery processing

- Export permit

- Goods issue procedure

- Transport handling

- Transport disposition
- Rebate procedure
- Credit-/debit memo processing
- Customer contract procedure
- Invoice processing
- Customer master handling

Questions concerning the Customer contract handling

Which business partners occur in the customer contracts?

- Ordering party
- Goods recipient
- Invoice recipient
- Payer
- Forwarding agent
- On which validity periods is the customer contract based?
- How are prices determined?
- Which factors can affect the pricing?
- Customer
- Material
- Amount
- Production costs...
- How is the availability check proceeded?
- Static
- Dynamic
- On finished-product level
- On assembly level
- With clearing
- Which texts elements are in the order processing?
- Header texts
- Position texts
- How is the credit limit check processed?
- Static

- Dynamic

Question concerning the delivery handling

- How is the delivery handling checked?
- Are deliveries generated out of delivery stock?
- Which business partners are there in the scope of the delivery procedure?
- Ordering party
- Goods recipient
- Forwarding agent
- Do customers accept partial deliveries?
- Is the availability checked during the delivery period?
- Are deliveries generated separately or in common?
- Which documents are applied, in order to terminate the delivery process?
- (Packing list, bill of loading…)
- Is the credit limit checked during delivery procedure?
- Is a quality check of the goods, to be delivered, processed?
- Is a commissioning made? If so, how?
- How does the packing ensue?
- Is a barcode applied?
- How are freight charges handled?
- Which transport documents are generated?
- Is the backlog procedure supported? If so, how?
- Why is a sales document blocked for delivery?
- Which information must be stated in a delivery document?
- What has to happen, if the delivery quantity exceeds the order amount?
- How are the times required for picking and packing, the loading time, the transportation lead-time and the transit time determined?
- Which texts shall be stated in delivery documents?
- Are following documents transferred?

- Delivery note
- EDI- messages
- Further more
- What are the criteria of export control?

Questions about the condition procedure

- Through which master data combination is the pricing supported?
- Which pricing information is used for statistic reasons?
- Are manual changes of prices processed?
- Is the pricing used by Interval scale?
- Is the value added tax included in prices?
- Are rebates offered to customers?
- If that is the case, on what do they base?
- Are these actions customer specific or do they depend on price?
- The pricing is based on which date?
- Date customer order
- Desired delivery date
- Valid due date
- How are the freight costs calculated and billed?
- What information about subtotals shall be determined in the pricing?
- Are prices calculated with the help of formula?
- Are external tax packages applied?

Questions about the Customer-Material-Info-Procedure

- Is it necessary, to check the material before transport to a customer?
- If yes, shall features or errors of a material be registered?
- Are quality certifications given to customer?
- Are quality info records used for customers and materials?

Questions about the customer master processing

- Who is responsible for the maintenance of customer master data?
- Central area
- Every area
- What customer groups are in a company?
- Customers
- One-time customers
- Interested party
- Commission recipient
- Rebate recipient...
- Are there deliverers, who are also customers?
- Do customers have various delivery sites and paying offices?
- Are customer specific calendars used?
- Which payment conditions shall be defined?
- Which are the Incoterms of the customers?
- What is the information about the contact persons of customers?
- Are sales information registered in the customer table?
- How are the customers classified?
- Is the sales analyzed referring the customer?
- On which customer groups shall the liabilities for the finance distribution be distributed?
- Inland
- Foreign countries
- End customer

Question concerning the invoice procedure

- How are invoices created? Separately or out of the invoice stock?
- Which business partners are there in the scope of invoice procedure?
- Ordering party
- Invoice recipient

- Payer
- Frequency of the invoice creation?
- Are there predefined times, when customers shall receive invoices?
- How are the invoice documents generated? Referring deliver documents, sales or other variables?
- Which invoice documents are created?
- Are pro forma invoices necessary?
- How are price changes recognized arising between customer order and invoice date?
- Because of what reasons is a delivery blocked by an invoice creation?
- What information is necessary in an invoice document?
- Which texts shall be given in an invoice?
- How is the text finding proceeded?
- What can influence the pricing?
- Customer
- Customer group
- Purchase quantity

Questions about the rebate procedure

- How many times are reimbursements made? Weekly, monthly…?
- How does the reimbursement ensue?
- How is a rebate calculated?
- Are partial billings on rebate processed?
- How is the percentage of customers receiving a rebate?

Questions concerning the contact processing in the scope of the sales support?

- Are customer calls or calls of potential customers registered?
- Is information about competitors or competitive products registered?
- Shall information about future customers be registered?
- Are customer complaints registered in the customer service?

- Are address lists for direct mailing used?

Questions about the credit-/debit memo requirement procedure?

- Do credit-/debit memos run through an approval process?

- Which adjustment types are processed?

- Which adjustment methods are applied? Free or bound to order/invoice?

Questions concerning the goods issue procedure?

- How is material taken off stock, registered?

- What time runs by, from the actual goods issue to the posting in a system?

- Is the posting processed online or in a collective run?

- Is the stock after the stock transfer posted to the receiving plant?

- Which documents are generated in connection with the goods issue posting?

- Which information is included in these documents?

Questions about the transport disposition-and procedure

- How many forwarding agents are appointed for the transport procedure?

- Do interfaces to the foreign transport systems exist?

- Does a customer have to pay the freight costs?

- How is the transport route determined?

- Are single or collective transports processed?

- Which transport methods are applied (truck, ship, rail...)?

- Are freight forwarders appointed?

Questions concerning the processing of the export permit inspection

- Does a definition exist about the legal regulations of the respective country?

- Is the export license customer specific?

- Is the export license sales transaction type specific?

- Is the export license product type specific?

- Is the export license limited by amount of money or a quantity?

- Shall the export license be checked at the delivery date, or either at the customer order and the delivery?

Questions about the customer order processing

- Are inquiries or quotations changed in customer orders?

- Which information is registered in the customer order?

- Which reasons are there for a customer order?

- Why can an order be rejected?

- Is it checked, whether order positions have already been used in another order?

- Are goods sold, which a company has bought from a deliverer?

- Are these materials always bought or only in case of an order?

- Are through delivery avis documents created for the delivery inspection?

- Is it checked, whether a customer has an open contract, by generating an order?

- Are order agreements sent? If yes, by which means?

- Which texts are to be used in order documents?

- Which information shall be displayed on the screen with the further processing of customer orders?

- Is a transit time included in a calculation of a delivery date?

- How is the availability referring an order date determined?

- How is the delivery specified at its planning?

- Which types of customer orders are there?

- Standard order

- Rush order

- Credit-and debit memo requirements...

- Is a customer charged with the amount transported by the customer?

- Is a customer permitted, to keep our material in his stock, while it remains in our property?

- Does the customer order material or is the transport regularly processed?

- Which method is applied, in order to determine, when a customer is charged with used material?

- How is one informed about when a customer has sold a product?

- Is a product, not needed or ordered, returned from a consignment stock?

- Is a customer permitted to return material in consignment stock, after its delivering?

12.9   Advantages of the ASAP-usage during the implementation of the R/3-module SD

Accelerated SAP optimizes time expenses, quality and efficient usage of all resources, in which all activities are coordinated, necessary for a successful termination of an R/3-project.

Summing up, the SAP-introduction method within implementation of the R/3-component SD-Sales results in following advantages:

- **Quicker R/3-implementation for a quicker profitability:** A target-orientated project management and a detailed project plan with predefined working steps and quality controls help keep R/3-projects clear. "The biggest danger is a creeping development of a project." "A quick implementation serves a quicker profitability." Such reactions come from companies of all sizes and sectors. Accelerated SAP offers strategies, for separating the essential from the unessential and proceeding a process orientated implementation.

- **Worldwide homogeneous implementation procedure:** It doesn't matter, whether R/3 is implemented within a company or in a foreign country in cooperation with other consulting-companies: all consultants have access to accelerated SAP and can use it for implementation.

- **Quality control and know how-transfer:** A quality control is added to every phase in accelerated SAP. In addition, ASAP offers evaluations, which evaluate the success of a project and display possible risk factors within an R/3-implementation. Beyond that, the SAP-consultants make their know how available during the entire realization of a project, in order to make sure, that the customer is independent after a successful introduction.

- **Efficient usage of existing resources**: It is a well-known fact that every project could do with more employees. The ones have other engagements, the others are employed in different sites. Accelerated SAP considers these circumstances and recommends, what background knowledge, experience and which other knowledge the members of a project team should have.

- **Reutilization of future implementations:** Documents as e.g. Business Blueprint, can be reused for a company-wide implementations or a new R/3-phase or, if required, adjusted and expanded.

- **Everything at one place:** Has a company already processed an R/3-project, it possibly has therefore many special versions of various accelerators. ASAP on the other hand provides as a starting point all parts completely at one place.

- **Standardized documentation of the R/3-solution:** ASAP contains a homogeneous concept for all types of documents, which are typically generated during the implementation process. As far as possible, the documents are reused. Therefore are, e.g. in the user documentation parts of the Business Blueprint applied.

- **Nothing is forgotten:** Accelerated SAP ensures that nothing is left out. It coordinates in particular the interaction between commercial and technical tasks: For our researches have shown that project managers normally require more support in one of the two areas mentioned- depending on the background they have.

- **Insights in implementation process:** With the accelerated SAP Roadmap it is easier than ever before, to duplicate the implementation process, particularly if persons are in the team, who have never before processed an R/3-introduction. It is possible, to understand everything without any problem, independent from where you are, what is planned, what the parts of every single project are and no matter what you do.

- **Concentration on former project states:** Accelerated SAP pays attention to various transactions, being easily underestimated and in the past being cared for too late, as e.g. conversion of data, definition of interfaces or allocation of access rights. The principle is, to determine problems in the phase of the Business Blueprint, so that they can be faced with regard of budget, time frame and resources available. That matters in particular for possible functional gaps.

12.10   Problems of ASAP

The ASAP-implementation method has to struggle with the same problems as the classic procedure model for an R/3-introduction, since it is also based on waterfall-based approach.

As you can recognize by taking into consideration the variety of working packages to be processed in the scope of the ASAP-procedure model, an R/3 implementation is afflicted with following problems:

- High time expenditure
- Resulting high financial expense
- Not fulfilling of various requirements

You can meet the problems that develop in the context of an ASAP implementation effectively with CATT. You can use CATT in each phase of an ASAP implementation and across various phases, including the following tasks:

- Automating the creation of master data
- Automating the transfer of master data from the legacy systems
- Automating the reading of data from spreadsheet programs
- Testing Customizing settings
- Executing module and integration tests
- Creating training data
- Training users with IDES
- System testing
- Checking error messages from the system
- Checking valuation and database updates
- Generating test data
- FCS version and release tests

Because the ASAP approach arose from the traditional SAP procedure model, both contain several identical work packages. As a result, many of the CATT applications that you use with ASAP are comparable to those of the traditional procedure model. However, you can also use CATT in many of the supplementary work packages of the various ASAP phases. As an example, we include the following work package as an example:

Work package: *including predefined settings*

The purpose of this task is to use configuration packages to accelerate the configuration process. It therefore involves testing, and, if necessary, changing, the predefined configuration settings.

ASAP includes predefined settings for a predefined client for the US market. This predefined client (PCC) provides a tested environment oriented to the US market. This client features:

* The starting point for configuration

* A basis client with simple structures upon which you can create and check configuration scenarios.

For example, you can use CATT to run through business scenarios for purchasing that form part of the PCC. Use of CATT in this context simplifies and accelerates learning the R/3 System. To reduce complexity, unnecessary adjustments are deleted.

Following process is recommended:

1.   Cooperation with the technical business process team

Here, the times and date for the loading of the predefined client.

2.   Cooperation with the business process team

Check, of whether the predefined applications for the system implementation are useful for the project teams. Further information about the predefined client is in this context found in the Knowledge Corner of the ASAP-tool. Experiences of many R/3-customers show, that 4-6 weeks of implementation can be spared, due to the usage of the predefined ASAP-client.

The installation period for the predefined client takes about 3 hours. This shall be considered before deciding, whether the predefined client is to be used or not.

3.   Installation of the predefined ASAP-client

See *predefined client* in the Knowledge Corner, in order to install the predefined client and receive corresponding information.

4.   Tools

Predefined ASAP-client

5.   Roles

Business process team leadership

Technical team leader

With regard to a precise description of the application of the Computer Aided Test Tools as integral element of a R/3-introduction, either with the traditional and the ASAP-implementation method, the reader may be referred to the book "Testing SAP R/3 Systems" from Gerhard Oberniedermaier and Marcus Geiss, published in the SAP Series by Addison-Wesley.

12.11 R/3-introduction tools for implementation

Already before the start of an implementation project it must be considered, what shall be achieved, which is the optimal introduction sequence and what business processes meet the needs of the own company the best. An R/3-introduction can, in this context, be supported by the already described Implementation Assistant, by the usage of following tools:

- R/3-Business Navigator
- External modeling tools
- Implementation guide (IMG)
- Concept Check Tool
- Project Estimator

12.11.1 Business Navigator

The reference model is completely integrated in the R/3-system by the Business Navigator. This tool enables direct access to application transactions, definitions in the ABAP-dictionary or project documentations. The Business Navigator differentiates between two views on the reference model:

- Component view
- Process view

In the reference model, the functionality of the R/3-system in the standard is described. Hereby following components are depicted:

- The process model: Procedure organization
- The structure model: Structure organization
- The distribution model: Presentation of distributed business processes and structures

- The business-object-model: object-orientated view on R/3 functions and data

The usage of one of the described models for illustration of company requirements is much more profitable than the introduction of a real system. It is possible to simulate and also company-related define different R/3-processes and function variants by the reference model.

12.11.2   External modeling tools

In the scope of an implementation of the R/3-system the necessity can arise, that the reference model is to be extended or changed for the documentation of specific requirements. For most companies it is necessary, to document requirements going beyond SAP-standard, in order to process a qualified deviation analysis by establishing models. These models can be created with different modeling tools and used as model for own development plans.

As examples for such modeling tools following are to mention:

- IntelliCorp Inc.: Live model

- IDS Scheer   : Aris Toolset

- Visio   : Business modeler

- Micrografix   : Enterprise Charter

12.11.3   Implementation guide

The implementation guide is a tool for customer specific adjustment of the R/3-system. It contains for every application component:

- All steps for introduction of the R/3-system

- All standard features

- All activities for definition of the R/3-system

The hierarchic the structure of the IMG forms the structure of the R/3-application, lists all documents relevant for implementation and includes active functions with following purpose:

- Start of customizing transactions

- Creation of project documentation

- Maintenance of state information

- Support in the control of implementation

There are four steps of an implementation guide:

The SAP-Reference-IMG:

It contains all documents for all functions delivered by SAP.

The Company-IMG:

This IMG is a part of the SAP-Reference-IMG, including only these documents of countries and functions to be implemented.

Project-IMG:

Project-IMGs are parts of the Company-IMGs, including only these documents of the Company-IMG, which shall be introduced within a customizing project.

Release-specific IMGs:

They are based on either the company-or project IMG. They show all documents for a certain release change, which are connected with release information.

The functionality is always the same, independent from the implementation guide being processed at the moment.

The IMG is consequently used for definition of all system parameters for the business processes in R/3. It contains functions for the project management and a component-oriented view on all customizing activities. Therefore it is possible, to document every activity in detail as well as allocate responsibilities and states.

12.11.4   Concept Test Tool

The Concept Test Tool is a tool, with which a quality check for project preparation, technical infrastructure and definition features can be processed. These checks are at first performed in the first two introduction phases. Therefore you are always informed about the potential conflicts referring data volumes and definition, which can develop to performance problems.

12.11.5   Project Estimator

The project estimator is an SAP-internal tool with which the consultants can exactly calculate required resources, costs, technical infrastructure and the time scale of implementation. The Project Estimator considers the project volume as well as different project-and risk factors.

An essential advantage of this tool is the standardizing of evaluation basis and therefore a much better comparability. Based upon that it is possible to process deviation analysis.

Consulting Books

Nikolaus Mohr/Gerhard P. Thomas
Interactive Broadband Media
A Guide for a Successful Take-Off
2001. xii, 177 pp. with 48 figs. Hardc. € 74,00   ISBN 3-528-05781-5
*„Das Buch bietet eine fundierte Analyse des Markts der Breitband-
medien. "*   Kress-Report 38/01

Andreas H. Schuler/Andreas Pfeifer
Efficient eReporting with SAP EC°
Strategic Direction and Implementation Guidelines
2001. x, 217 pp. with 112 figs. (Efficient Business Computing, ed. by
Fedtke, Stephen) Hardc. € 99,00   ISBN 3-528-05761-0
Contents: Management reporting and legal consolidation Intgration of
management reporting and legal consolidation and integration roadmap
- Reporting and the role of information technology - Efficient Reporting
with SAP R/3 ES - Design, Implementation, Test, and Deploy -
eReporting

Heinz-Dieter Knoell/Lukas W. H. Kuehl/ Roland W. A. Kuehl/
Robert Moreton
**Optimising Business Performance
with Standard Software Systems**
How to reorganise Workflows by Chance of Implementing new
ERP-Systems (SAP, BAAN, Peoplesoft, Navision, ...) or new Releases
2001. xvi, 425 pp. with 191 figs. Hardc. € 74,00   ISBN 3-528-05765-3

Abraham-Lincoln-Straße 46
65189 Wiesbaden
Fax 0611.7878-400
www.vieweg.de

vieweg

Stand 15.3.2002. Änderungen vorbehalten.
Erhältlich im Buchhandel oder im Verlag.

13 Supply chain management

In the passing years supply chain management (SCM) has grown an important concept of company management, which is paid attention to. In the scope of this new management concept, the idea of the Lean Management is spread over the entire value added chain. Herewith the elimination of company internal extravagancies along the value added chain is strived for. SCM can have a variety of functions, which shall now be presented in a brief survey.

13.1 Functions of SCM

Subsequently, a short survey about the objectives of SCM is given in extracts [Halusa 96/23]. Following targets are fundamentally differentiated:

- Strategic targets
- Operative targets

13.1.1 Strategic target

- Development of strategies for product and process development
- Development of strategies for goods and services
- Quality management
- Procurement-and sales strategies
- Distribution strategies
- Deliverer-and customer management
- Legal performance of co-operations
- Joint search for better business processes

13.1.2   **Operative targets**

- Optimization of internal logistics functions as transport, storage, order quantity-and lot size optimization

- Procedure optimization within business processes

- Introduction of internal information processing systems for planning, control and inspection of the order lead

- Sales-and procurement market research

- Field service control

- Mastering of organizational and system technical interfaces

Now, the essential functions of the order procedure for the area of sales being contacted by SCM shall be presented. The possible application and procedure type is added to every function described. Companies must analyze eminent data amounts of most different sources, in order to be able to make substantiated decisions to the logistics chain referring functions and requirements of SCM. Many of these data have their origin in the business processes of the company itself; of deliverers, partners and even customers. In contrast to the data models of existing ERP-systems, the systems for the decision support of logistics functions require a new data model, which can proceed an enormous quantity of complex data in real time. Up to now, enterprises had to face the problem, of connecting software solutions with their underlying ERP-system by themselves and of also realizing special interfaces, which could import data of external sources. This, however, took a long time in practice and required expensive implementation cycles. In consequence it is: To optimally provide the logistics chain with information, it is necessary to connect the internally applied ERP-systems with a qualified SCM-software. The SAP AG has therefore initiated the so-called "New Dimension Initiative" for the subject SCM and developed an own SCM-software. This SCM-software is also described in the further course of this book.

13.2   Supply chain management in sales

Many approaches of SCM are based upon questionnaires of logistics. In this context there are also conceptions, which are essentially focused on sales. Exemplary areas of sales, affected by SCM, are:

- Order processing
- Creditworthiness check
- Time inspection
- E-commerce

13.2.1   Order processing

The processing of necessary external data for an integrated information procedure is the target of order processing. With the SAP products, different types of order processing are realizable:

- Transmission by EDI (Electronic Data Interchanges): Herewith, information can be transferred from a foreign system to the R/3-sales module SD

- Connection of two SAP systems by ALE (Application Link Enabling)

- Order entry by the customer via Internet: Hereto it is necessary, to apply the SAP Internet Transaction Server (IST) and the Internet Application Components (IAC s)

- Complete order processing through a field service employee via laptop

In the scope of these prewritten different types of order processing within SCM, an order and technical check should be integrated.

13.2.2   Creditworthiness check

With every order processing a creditworthiness check module should activated. Herewith it is important, that the check is not statically but dynamically made. With a dynamic check, dates for the payment receipts are forecasted. Therefore a possible exceeding overdraft, caused by orders just received and checked, can be recognized. A limit for an amount of open claims and of orders not yet billed, is in general be stored in a customer master record. This requirement is realized in R/3 through the credit management in the modules SD and FI.

13.2.3 Deadline check

During the order processing and-inspection, it is necessary to examine, whether the customer desired delivery date stated in the order, can be met. An adherence to the desired delivery date can be of great importance for customer ties. This can be realized by usage of a deadline check module. This module processes the so-called "ATP (available to promise)-check". Hereby it is checked, whether a delivery promise can be given or met. Following questionnaires are considered:

* Is it possible to meet a delivery obligation referring amount and date?

* If yes, which date can be promised to the customer?

* Can reservation be touched?

* Is the check limited to a warehouse and/or production site?

The R/3-system is able to calculate availability dates on basis of available warehouse stocks, subsequent deliveries, capacities for self production as well as duration for foreign purchase. The most efficient and differentiated SAP-product for the deadline check is at present the module ATP-Global availability check of the New Dimension product APO-Advanced Planned Optimizer.

13.2.4 E-commerce

Relating to E-commerce, two main scenarios are registered:

1. Business to business (B2B): For the processing of scenario B2B, during which a company orders, receives an invoice and also pays via net, SAP offers the New Dimension-product "Business to Business Procurement". This product allows performing business relationships on basis of Internet, in the sense of SCM. As already mentioned, the B2B scenario can control and process business transactions outgoing from order requirement over inquiry up to the payment of an invoice.

2. Business to Consumer (B2C): This scenario is also called "Electronic Retailing". In so-called shopping malls many types of consumer goods can be offered to customers in

the Internet. This scenario can be displayed with the help of the R/3-component SAP Online Store.

Now, the realization of these strategies in the SAP environment shall be examined. The SAP AG has, referring the supply chain management- initiative, which belongs to the New Dimension-initiative of SAP, developed different tools:

- APO-Advanced Planned Optimizer
- Business to Business-Procurement
- Logistics Execution System

13.3   APO- Advanced Planned Optimizer in sales

13.3.1   Introduction

Companies having already changed the structure of their business processes with great expenditures must now look beyond their self defined borders, in order to gain a permanently continuing improvement.

Companies, which up to now have not performed an optimization of their business processes, must develop strategies referring market requirements and therefore also employees demands, in order to effectively improve their competition capability.

For this purpose it is necessary, to build up a new IT-infrastructure which enables the immediate reaction towards permanently changing market conditions. Customer satisfaction shall herewith be defined prime target and consequently the competition capability and profitability secured.

Because of this, SAP has developed the initiative "Supply Chain Optimization Planning and Execution" (SCOPE). SAP SCOPE guarantees a connection between the reliable R/3-system, business data in the Business Information Warehouse and an efficient analysis system.

In today's ERP-environment also more and more persons are incorporated. That means, that ever more persons take part in electronic processes of a company. Dialogue between sales employees, planners, deliverers and customer gains ever more importance for the company success. By means of SAP SCOPE these information are far stronger integrated in business processes than before. Data and decisions of the entire logistics chain can herewith be integrated in a unique automated performance environment.

Essential element of SCOPE is APO. It improves the information flow and integrates not only team orientated decision finding and progressive planning in real time, but also optimization of the R/3-system.

13.3.2   APO-survey

APO includes planning functions for the strategic, tactical and operational planning of logistics chains. Integrated APO-modules cover the entire field of planning. APO is provided with several integrated modules, which have access to a joint data basis. In the following, these modules are briefly introduced and described.

Supply Chain Cockpit

Supply Chain Cockpit (SCC) is a graphical instrumental board for modeling, performance, planning and control of the supply chain. With this tool the user can on the one hand have view of the logistics chain as such, and on the other hand it shows control possibilities of underlying planning-and scheduling processes.

The Supply Chain Cockpit contains the Supply Chain Engineer. With the help of this tool, it is possible to create specified graphical representations of the logistics chain. This is structured by single components (also called network nodes), which are included in a library. For example sales warehouses, customer plants and transport routes are among these components.

If the logistics chain is composed, a series of real time releases can be connected with single network nodes in the logistics chain. Through this release the APO can be adjusted to the connected R/3-system and to systems of external producers.

Typical users of the SCC can be:

- Planners, building or modifying a new delivery network

- Demand planners, having access to requirement forecasts

- Sales planners, creating and changing on basis of delivered details concerning single network nodes

- Controller, analyzing elementary data and ratios and looking for possibilities to improve logistics structures and processes [Knollmayer 99/107].

Demand –and sales planning

The capability of precisely forecasting the demands is very important for the company success. In view of the permanently growing customer expectations and the competition becoming ever more global, the demand is subject to a high fluctuation and breadth. It is therefore the prime target of APO, to meet these demand fluctuations. This takes place by means of support through the APO-module DP (Demand Planning). DP helps to forecast the demand for saleable products. The forecast support with DP shall lead, in comparison to present procedures, to an evident improvement of the forecast quality. The complete integration of the APO- sales planning offers following advantages for users:

- Influence of activities of the logistics chain in real time

- Optimized usage of stocks by means of exact medium-term forecasts

- Support of the strategic planning by long-term forecasts

Typical usages of the sales planning are:

Joint forecast performance

Harmonized forecasts for the demand planning can be performed of the business fields marketing, sales/distribution and even deliverers. These forecasts can be actualized in real time and display problems to be expected.

Life cycle administration

Here, the time planning for product introductions are determined. Factors, as superior products, replacement and displacement are considered. Consequently, administration of the product life cycle is placed on a stable basis.

Special offer planning

Unforeseen demand fluctuations can be balanced by activities like special offers. APO serves the planning of such activities. Factors as profit margins, product availability and historical trends can be considered. By means of past data, effects of price markups or-cuts can be determined for future demand.

Forecasts for new products

APO supports the creation of forecasts for new products through deduction of models off forecasts and demand history of similar products. Product introductions can be supervised e.g. by POS-data.

Causal research

Effects of external factors on a company success are inspected. Thereto belong:

- Demographic changes

- Environment conditions

- Social factors

- Political factors

Even future demands can be recognized and forecasted by means of the APO-sales planning.

Supply Network Planning, Deployment and Transport Load Builder

With the APO module Supply Network Planning (SNP), a planning method for a creation of tactic logistics network is made available to enterprises. With the help of Deployment and Transport Load Builder, distribution network, and optimal usage of transport means can be planned. The companies are enabled to exactly coordinate offer and demand. Therefore, APO integrates the subsequently mentioned business fields in a consistent model:

- Purchase

- Production

- Sales/Distribution

- Sales and

- Transport

Through this modeling of logistics network, all logistical activities can be coordinated and the material flow optimally planned along the whole logistics chain. Realizable plans for purchase, production, inventory management and transport.

APO includes also functions, with which companies can dynamically establish, when and how the stock shall be distributed. Through this combination of logistics planning and distribution, the optimal usage of production-, sales and transport resources for the covering of the forecasted and actual stock is secured. Herewith, following targets, which are also prime logistics targets, can be realized:

- Customer service improved

- Reaction time shortened

- Reduced stocks and buffer stocks

- Shorter cycle periods

- Higher profitability

- Lower logistics costs

Following functions are supported by the APO-modules SNP and development:

- Complete coordination of the logistics chain with the help of different algorithms (e.g. heuristic, linear programming...)

- Planning, creation and optimization of individual data plans and surveys. Hereby, e.g. transport, warehouse capacity and so forth are considered

- Simulation of strategic network planning. Consequently, deliverers, plants or sales stocks can be added

- Offer and demand can be dynamically coordinated

- Allocation recommendations and-strategies can be developed

- Realization of a short-term stock and distribution optimization: Here, the determination of the optimal distribution of an available offer as reaction to a short-time demand, takes place

Example for illustration

In the SD-module and the APO a total demand plan is established after the order receipt, which does not consider any capacity and material limitations. Subsequently a medium-term production-and distribution plan comes into existence with the help of methods integrated in the Supply Network Planning. These plans can be interactively adjusted to circumstances, which up to now were not considered or through the usage of Heuristics. The resulting plan is made available to the APO-modules production planning and detailed planning. Here, the production plan is coordinated with the actual state of the production system and a realizable plan with consideration of possible bottlenecks is established.

After the usage of SNP, the deployment method checks, which demands are to be covered by planned supply activities. If offer and demand correspond with actual values, the plan is accepted by the Deployment. Otherwise the plan is adjusted with different distribution rules. In the subsequent step, accepted orders are transmitted to the transport plans with help of the Transport Load Builder [Knollmayer99/124ff].

Production planning and detailed planning

To be able to quickly react to market requirements, not only exact demand forecasts, but also capabilities to cover the demands are necessary. The APO-module Supply Chain Network Planning and Deployment allows, that the material-and resource flow optimally runs through the entire logistics chain without clashes. The module PP makes optimization techniques for the short-term material-and production planning available, by taking into consideration capacity fluctuations and enables this therefore for every single plant. The detailed scheduling DS helps to allocate production resources and the sequencing. The result is a practicable production plan.

Following advantages for companies result:

- Higher customer satisfaction through avoiding delays

- Higher profitability through better usage of bottleneck resources

- Detailed capacity planning simultaneously with the material planning

- Multiphase forward-and backward scheduling becomes possible

- Multiphase transmission and adjustment of changes can be realized

- Backlog orders can be integrated with rescheduling

- Lower overtimes on grounds of practicable production plans

- Lower stock costs and WIP inventory, because of material releases.

An example shall clarify the integration of production-and detailed planning in R/3 and the resulting advantages:

A customer places an order via Internet. This order enters the R/3-system (SD). An ATP-requirement for the product is created out of the SD-module to APO. With the help of APO it is determined that this product must be produced in an individual production. The customer order passes now the model of production-and detailed planning. In addition a promise date for the product to be produced is determined and forwarded to the SD-module.

Global availability check

A further basic function of APO is the "global availability check". This function offers the possibility to equalize offer and demand with the help of a rule-based strategy. In this connection, multiphase components-and capacity checks in real time or as simulation are performed. As a result, this function initiates an immediate and simultaneous access to all availability data of products beyond the entire logistics chain. This can be of great importance for a company, since therefore e.g. the customer tie is increased or penalties can be avoided, which would become due, if delivery dates fixed in contracts cannot be fulfilled.

13.4   Business to Business Procurement

The application B2B procurement is one of the first applications being delivered by means of the SAP initiative Enjoy SAP. It enables procurement of C-materials, which are not planned by an ERP-system, like office supplies, operating supplies and services. The SAP B2B procurement allows therefore an open, complete business overlapping procurement and controls all procurement processes from order requirement to invoice payment. With the help of the electronic catalog the user can, by most simple means, directly procure goods and service from his/her place of work. The purchase department is thereby facilitated and can concentrate on the strategic purchase as e.g. on the component supply management or on price negotiations. Use of this efficient software optimizes the internal order requirement as well as indirect procurement processes and reduces by it the total costs in the purchase to a great extent. This B2B procurement solution for the procurement of operating supplies, goods and services organizes and automates the transfer of procurement transaction to the place of work of the single employee. Thereby order exactness is increased and order duration shortened. This benefits the result of the entire logistics. Following advantages arise thanks to the usage of B2B procurement:

- Incorporation of the employees by user friendly self-service functions

- Incorporation of electronic catalogs over open catalog interfaces

- Topical integration of purchase-and sales department

- Strategic deliverer management

- Voluminous evaluation functions

For procurement by means of B2B procurement, there are two possibilities

- Internet based, independent procurement

- Catalog incorporation

13.4.1   Internet based, independent incorporation

B2B procurement offers, on basis of the ever expanding Internet technology, the possibility for interaction between provider and purchaser. An employee being vest with corresponding powers, can consequently, by a mere mouse-click directly from his/her place of work, purchase goods and services required. To the activity range of a comprehensive procurement solution belong:

- Creation and maintenance of order requirements with or without catalog, that is, companies can charge their employees with the creation and maintenance of order requirement and orders

- Approval and rejection of order requirements in need of agreement: The Internet capable work flow functions of SAP R/3 allow effective control and inspection of the procurement process

- Goods receipt at place of work: This leads to an additional tightening of the procurement process, since it allows the reduction of overhead cost material of the field operational supply by the personnel.

13.4.2 Catalog integration

The SAP B2B procurement is also characterized by the functionality of flexible catalog integration. Several solutions of the catalog integration are hereby supported:

- The administration of the company own catalog

- The integration of catalogs of different providers

- The direct access to the catalogs of providers

The open catalog interface ensures that all catalogs of external providers can be connected to the B2B procurement. Because of the direct catalog access, a B2B interaction is built up between provider and purchaser. If on part of the deliverer the SAP-Online Store is applied, the B2B procurement generates a customer order at the deliverer and in the purchase department of the customer an order [Knollmayer99/138].

13.5 Logistics Execution System

The SAP Logistics Execution System integrates warehouse management and transport management functionalities with other SAP supply chain solutions and R/3 business processes in an overlapping supply chain solution.

13.5.1 Warehouse management

The warehouse management system is a component of the logistics executive system. Thus, it can also be applied as an independent component. A warehouse management system supports, among others, following functions:

- Administration of warehouse structures and equipment

- Survey of stock movements

- Control of stock data on bin location level
- Transfer and release from stock of dangerous goods

Based upon data of the Logistics Information System (LIS) of R/3 rough- as well as detailed planning can be made.

13.5.2   Transport management

Route planning and transport costs management can be made on grounds of the transport management. This component of the Logistics Executive System comprises functionalities to the:

- Dispatch scheduling
- Route planning
- Freight cost calculation
- Transport procedure-and supervision

14 Support of implementation of SAP R/3 with Computer Aided Test Tool (CATT)

14.1   Starting situation

Enterprises understandably want to keep implementation times for new systems and new processes short. Nonetheless, they also want the systems to feature the highest possible quality and productivity. SAP reacted early to this demanding challenge. It made effective methods and tools available that helped its' customers reduce implementation times sharply.

In this context, SAP offers its proprietary *Computer Aided Test Tool* (CATT): a universally applicable tool. CATT is particularly well suited to support and accelerate development, modification, maintenance, and later additional development of R/3 Systems broadly and efficiently. For example, the tool can automate modular business processes of any kind and repeat them as often as desired. In addition, CATT contains all functions needed to start, test, administer, and log both individual transactions and test procedures for dynamic business or administrative processes.

The target of this chapter is, to give a brief view of application CATT with implementation of the R/3-component SD - SALES AND DISTRIBUTION referring the popular procedure models (traditional procedure model, ASAP-Implementation methodology, DSDM-based procedure model). Thus, the R/3-consultant as well as the user are provided with a practice orientated guide, which describes in detail the application possibilities of CATT within the introduction and also the optimization and further development of the SD- module.

Extending and searching information to the subject of this chapter are to be found in following books of the "SAP-Series", edited by Addison-Wesley-Longman:

„Testing SAP R/3 Systems", by Gerhard Oberniedermaier and Marcus Geiss, ISBN 0-201-67517-X. This book is a detailed practical guide for all R/3 consultants und ABAP programmers to the use of CATT, SAP's computer aided test tool, during the implementation of SAP R/3. It tells how to automate tests and optimize R/3 business processes, use CATT tools at each stage of an R/3 implementation, and how to use CATT tools in tasks such as testing transactions, testing database field values, configuring system tables, and generating test and training data.

„Dynamic Implementation of R/3" by Marcus Geiss and Roland Soltysiak, ISBN 0201674831. In this book, the authors present a new dynamic implementation method, which has been tested and proven to avoid the common pitfalls associated with the standard SAP procedural model. Following the principles of the Dynamic Systems Development Method (DSDM), the authors explain how to overcome these problems via a requirements-based approach and a new procedural model, which their experience has shown to be versatile, collaborative and, above all, highly effective. The book takes a step-by-step look at this model, detailing the specific activities, which should be initiated at each phase and showing how best to execute them. It explains how the model works in an iterative and incremental way, building on essential cooperation between IT departments and user areas. Central to the whole cycle is the close attention to quality management and efficient use of the Computer Aided Test Tool (CATT) on individual modules and processes, thereby supporting integration and promoting reliability.

14.2   CATT as quality assurance tool within the SD-implementation

The development of software, which fulfills highest quality requirements as e.g. the in Europe leading quality standard referring the software development, ISO 9000-3, has to be accompanied by solid, intensive tests.

Quality assurance and software tests – for many DV-people in the past, these duties were rather annoying. Since the ISO-9000-certification this view has fundamentally changed. Now, the software is generally accepted as important part either of a software development project and of a software-introduction phase. In both cases the test phase plays a considerable role within the project planning.

The target is always, to test with as few time-and personnel expense as possible, all eventualities in the fields functionality, stability and performance. In order to efficiently reach this target, special tools are applied, enabling an automation of the procedures.

With such extensive software as R/3, the quality assurance is not to be neglected at all. Even from the start SAP tested and examined the quality of the software with a proprietary, in the first only internally used tool with the name CATT (Computer Aided Test Tool). Since the version 3.0, CATT is an integral part of the system and is, as central test tool, available in SAP-projects. Since then, it has been permanently further developed and optimized. Today, it is as an integral part of every new R/3-release, already adjusted to its functions.

R/3 projects must place special emphasis on the comprehensive quality assurance. It's almost a given today that enterprises from various industries have certification according to a uniform quality standard, such as DIN ISO 9001. An enterprise that seeks certification according to this standard must expressly prove that its R/3 implementation meets the high demands of the standard.

A quality assurance plan therefore becomes one of the most important instruments for the effective quality assurance of an R/3 implementation project. In Europe, DIN ISO 9000-3 is the most important standard for quality assurance in the development of software. In three parts, the standard describes quality management for the development, delivery, and maintenance of software. The three parts include:

- Framework
- Life cycle activities
- Cross-phase activities

14.2.1   Framework of quality assurance

The framework describes now responsibility for the quality of software lies at the highest levels of leadership for the customer and the vendor.

14.2.2   Life cycle activities of quality assurance

The *life cycle activities* describe individual aspects of quality assurance. These activities include:

- Determining the requirements
- Planning development
- Planning individual phases
- Testing the software

14.2.3   Cross-phase activities of quality assurance

The *cross-phase activities* include aspects that can significantly affect all phases of a project, such as control of documents, rules, tools, and training.

As noted, individual tests are performed in individual phases within the software life cycle. To meet the leading quality standard for software development, DIN ISO 9000-3, in the implementation of R/3 software optimally, the activities described by DIN ISO 9000-3 as part of life cycle activities form a very important and unavoidable project activity. SAP supports this testing with its proprietary testing tool, the Computer Aided Test Tool (CATT) and the Test Workbench.

14.3   Definition CATT

CATT contains all necessary functions for the performance, start, test, administration and protocol of test components for single transactions, as well as for test procedures for dynamic operational or administrative processes within the R/3-system.

These operational processes can on the one hand be limited to single transactions respectively to an R/3-module (functional orientation); certainly can on the other hand also operational processes be recorded, which extend over several R/3-applications (process orientation).

The usage of CATT reduces on grounds of this automation of test procedures the amount of expensive manual tests to a great extent and forces beyond that to the systematizing of different tests through an exactly defined input with a planned test result. Thereby detailed conclusions can be drawn referring the quality of the system. Moreover, manual test activities during the actual test period can be considerably limited, by the usage of CATT. With the general availability of an R/3-version, tests can be processed by CATT-functions within a few days.

14.4 Usage fields of CATT with the implementation of the SAP R/3-module SD

During the implementation of the sales and distribution module, you can use CATT as follows:

- Testing transactions
- Structuring master data in the context of transferring old data and creating new master data
- Setting up Customizing tables and testing the effects of Customizing settings
- Executing a closing test
- Executing module and integration tests based upon the mapping of business process chains with CATT
- Setup of training data
- User training with IDES
- System testing
- Check system messages and error messages
- Check valuation and database updates
- Generate test data
- FCS version and release regression tests
- Tests and checks of interface programs
- Connection of SAP business workflow and CATT is possible

14.4.1   Testing Transactions

To a large extent, the various processes of the R/3 System consist of numerous individual transactions. A transaction is a logically closed operation in the R/3 System. From the user's viewpoint, a transaction represents a unit, such as the creation of a list of specific customers, changing a customer's address, and so on. From the viewpoint of dialog programming, transactions represent complex objects, called by a transaction code, that consist of a module pool and various dynamic (screen) programs. Once a user logs on to the R/3 System, three levels are distinguished from each other: the SAP level, the application level, and the task level. A transaction is an application that takes place at the task level. To arrive at the initial screen of an application, a user can navigate through the menu hierarchy or enter a four-digit transaction code in the command line. Use of a transaction code spares users the need to navigate through the various menus and takes them directly to the initial screen.

Transactions are programmed by SAP itself (before delivery of an R/3 System to a customer), or are created by customers themselves from scratch or by changing transactions to reflect customer requirements. CATT can test both standard transactions and customer-specific, modified transactions during the development process. When you upgrade an R/3 System by copying a new, higher release level, you can retest any transactions. When you copy an R/3 System from a development client into a test client, you must determine if the transaction can run on the new system. For that reason, you must retest individual transactions. Instead of going through all the transactions manually, by using the enter key on screen after screen, you can automate the process with test runs, called CATTs in the following. CATTs significantly reduce the efforts that users must expend to complete testing. The following cause errors generated by the system during testing by hand or with CATTs:

a) Incorrect installation of the R/3 System

b) A problem with the transaction

Once you have excluded incorrect installation of the system as the cause of the error, you must inform the person who created the transaction. The programmer holds responsibility for reworking the transaction.

14.4.2   **Structuring master data in the context of transferring old data and creating new master data**

When an enterprise replaces various legacy systems with an R/3 System, it must, of course, transfer all the significant old data, such as materials, vendors, and customers, into the new R/3 System to maintain the company's business.

Maintaining each data record manually in the R/3 System would be a complicated and involved effort. CATT provides the user with an automatic way to read and process external files, such as Excel spreadsheets and data records, in the R/3 System. This feature reduces the effort needed to transfer old data significantly, as shown by use of the tool in practice. As a precondition, however, the data must have already undergone detailed and reliable formatting in a spreadsheet program and exist in a form readable by CATT.

You can also use CATT to generate new master data of any kind in the R/3 System. If you need 100 materials for an integration test, you simply let the CATT procedure for *create material* run 100 times. Compared to the effort required to create 100 materials manually, use of CATT simplifies the task greatly.

This type of data transfer also provides a great deal of data security. In this context, data security means that the tool simulates manual entry, so that just as in batch input, the system runs through all the entry templates and executes integrity checks.

14.4.3   **Setting up customizing tables and testing the effects of customizing settings**

The *detailing and implementation* phase executes company-specific adjustments to the R/3 System with Customizing. You can use CATT to make changes to Customizing tables. A change stores the original values in an internal table; CATT can later restore the original values. This feature allows you to inspect the effects of changes easily, without having to worry about the original entries. CATT procedures can test the effects of the Customizing changes and, if needed, reset the changes back to the original status.

14.4.4   **Executing a closing test**

The phase activity *executing closing test* is an activity, performing one of the most extensive duties in the field of the tests. The closing test has the task, to prove that the requirements stated in the requirement definition are met. The test cases determined there, are lying behind the closing test. Not only the special departments, but also all positions back-or forwarded, participate in the closing test. In the scope of the closing test proof for the correctness of the application system is to be made. Herewith the application system is made valid.

14.4.5   **Executing module and integration tests based upon the mapping of business process chains with CATT**

With the R/3-introduction the detailed examination of whether single dependencies on business processes harmonize with the value added chain, is part of the integration test. The integration test includes also interfaces, edition-and print functions as well as system expansions. Further more, it comprises department overlapping areas and relies in this context on the results of preceding tasks. The test is processed in the quality assurance system.

As noted the SAP R/3 System consists of diverse, individual modules. In other words, the R/3 System of each SAP customer represents an integrative concatenation of several modules. The SAP R/3 System also displays a broad range of functions.

The integration of applications provides the basis for the availability of all functions within the system and, therefore, throughout the enterprise. A guarantee of such availability arises only when each module is tailored to the requirements of the enterprise and then undergoes testing module by module. Testing at the module level represents the first and lowest level of computer-supported tests in software development. You first test all the elementary modules as the foundation of a system of programs and only then test the integration of the modules. As a minimal requirement for testing a software module, you must run through all branches of the program once and ensure that the module reactions according to its specifications. A test of a module is also known as a single test. Users can enjoy support from CATT during execution of module tests. As an example, consider the Sales and Distribution (SD) module. In sales and distribution, a company first sells products or provides various services to a wide variety of business partners. After completion of the sale, a company fulfills distribution tasks such as shipping, invoicing, determining conditions, transport, and export. The R/3 SD application uses the data on products, services, and business partners as the basis for processing its tasks.

To create a customer order in SD, you must first have a customer inquiry, to which you respond with an offer or quotation. The system generates a unique document number for each of the documents involved: inquiry, quotation, and order). To create a quotation in the R/3 System, the software requires you to enter the document number of the corresponding customer inquiry. In this case, the document number enables the system to transfer all the data from the inquiry to the quotation automatically. Creation of the order requires entry of the document number from the quotation.

During the module test, CATT enables you to map the functional business process of *create inquiry, create quotation, create order,* and *create delivery and transport.* It also enables you to test the ability of the process to run without problems throughout the system. In this example, the module test run successfully when the customer has received the correct quantity of items at the right time and at the right time, according to the order. A successful test means that the process has run correctly. An absolute precondition for success is error-free transfer of data between each transaction. A purely functional module test for SD also requires that the appropriate mater data for materials, customers, and so on already exists in the system.

For whatever reason, if the master data does not exist, you can use CATT to generate the required master data as a basis for the module test.

14.4.6   Setup of training data

You can use CATT to prepare user training. For example, if you need to post materials for user training, CATT can automatically create materials and sequential material numbers for the trainees. You can also record and explain complete, complex business processes that extend across several R/3 modules. This feature saves a great deal of time: you record the process only once, and it becomes available for as many applications as you desire. As an option, CATT allows you to run test procedures step by step, and to display the logical sequence of the procedure screen by screen on the monitor. With this option, later users can display and understand the origin and further use of their data in the context of the integration offered by the R/3 System. You can also make Customizing settings with the procedures prepared in advance.

14.4.7   Support and processing of user trainings

Purpose of this task is, to prepare training documents for users per CATT-support. This can also include slides, graphics, texts, polices, screen material, company specific exercises... This material can be used for introduction trainings of the users and for trainings of new employees.

14.4.8   User training with IDES

SAP also uses the Computer Aided Test Tool in IDES (International Demonstration and Education System), an integrated and preconfigured training system. CATT handles the creation of master data, testing process chains, and the creation of training data.

To clarify matters, we present a brief discussion of IDES. IDES depicts a sample firm, an international group complete with subsidiaries in various countries. As such it offers itself as an outstanding demonstration guide that users can employ for self-study of the broad range of functions available in the R/3 System. IDES contains sample application data for various business processes in the R/3 System. The processes in IDES are mapped just as in a real enterprise: the processes include realistic characteristics. The feature clarifies the multifaceted functions of the R/3 System with easily understood examples. IDES does not focus on the depiction of individual functions, but on general business processes and their integration. With IDES, the user comes to understand how the R/3 System supports business processes and procedures.

IDES can represent and even deepen all the examples and exercises in SAP training programs. IDES depicts numerous core processes that represent a high level of integration and important business processes.

The make-to-order area of logistics, for example, contains the process *make-to-order production with a configurable product.*

The example covers production of a motorcycle and includes the following process steps in an R/3 System:

- Creating the customer order in the R/3 SD (Sales and Distribution) module

- Transferring the requirement

- Creating a production order and fulfilling it in the R/3 PP (Production Planning) module

- Shipping and invoicing in the SD module

- Monitoring of costs and revenue in the CO (Controlling) module

To run this process through the IDES system, the required materials must exist in the system. If the system does indicate the presence of no materials or insufficient quantities, you must create new materials. A CATT procedure can generate the preconditions (*create material*) for this process. An automated CATT procedure updates the stocks of individual materials to the quantity required for the process procedure. Each time that the material quantity then falls below the required level, users in IDES training run the *create material* CATT to generate new material. CATT significantly reduces the effort required to create various types of master data for IDES training. You can also use the Computer Aided Test Tool to test and automate complete IDES processes.

IDES is available in each upgrade release. Although it maps an upgrade to a new release, you can load the hot packages for each upgrade release. A new IDES release means reinstallation. This limitation is the only way to ensure delivery of tested, high-quality new master data, transaction data, Customizing, and documentation.

For more detailed information on this topic, please consult the SAP training materials on IDES and the online documentation in the R/3 System.

14.4.9   System test

Implementation of an R/3 System normally involves a double installation. The enterprise has both a test system and a production system.

The use of a test system and a production system means that when production begins, various authorizations, authorization profiles, reports, and old data are transferred from the test system into the production system. You can, of course, also transfer the CATT test modules and procedures that exist in the test system into the production system. After transfer of the required data, you use CATT to retest the production system and the test settings. If the settings in the test system were transferred correctly along with the master data, a CATT test in the production system will also run successfully. If this is not the case, you must recheck and, if necessary, correct the system settings. The introductory chapter of this book treats the system test. The final result of preparation for productive operation should result in a tested, error-free production system.

As delivered to SAP customers, an R/3 System already contains a large number of preset CATT procedures programmed by SAP. Customers can use them immediately for various purposes. These procedures enhance the Computer Aided Test Tool and make it easier for customers to use CATT. With its development partners, SAP has developed *industry models* that you can also use with CATT procedures to improve quality. Here, too, SAP intends to deliver preset CATT procedures to its customers.

14.4.10   Checking system messages and error messages

Transactions change data; they use system messages and error messages to confirm the success or failure of a procedure. When users enter values, transactions check the values for correctness. Examples include entry of a data in a specific format. If the entry is missing or incorrect, a system message or an error message appears. You can use CATT to determine if transactions contain the error or system message and if they react to the messages correctly.

You can also use system messages within CATT procedures to trigger specific actions. If an error message appears during creation of standard order and states, "customer does not exist," the corresponding CATT procedure can create the customer.

14.4.11   Checking valuation and database updates

With CATT, you can read values from database fields and thereby check database updates after transactions have executed. You can also examine a table for the existence of a particular entry. If you used CATT to create a vendor, you can check to see if the vendor actually exists in the appropriate table.

14.4.12   Generating test data

You should test transactions in the test system, not in the production system. To work in the test system, however, you must have data available. CATT can create the test data, such as materials, customers, vendors, and so on. Test data should be set up to ensure that it can consider all the possibilities that can occur when testing the data, such as material subject to handling in batches, language dependencies, and so on. These CATTs are also considered preparatory procedures. You can use them to make complete Customizing settings.

14.4.13   FCS version and release regression tests

As soon as an R/3 release exists in a FCS (First Customer Shipment) version, you can begin to work with CATT. Up to the general availability of the system on the market, you can prepare and implement any desired test activities that deal with the (numerous) new functions of the new release. You can also use these tests to develop a regression test packet dynamically, a packet that focuses on the decisive business processes for a particular SAP customer. A regression test packet ensures the ability of business processes and transactions already tested in earlier versions to run in the current version of the R/3 System. Accordingly, the use of CATT here during the testing period can significantly limit the scope of manual tests.

Practical experience has shown the advisability of installing all new R/3 releases in general availability immediately and subjecting them to the widest possible testing. As an unavoidable precondition for these tests, you must first create a flexible system environment. You should examine new releases simultaneously on several systems with an almost identical testing scope. You can use CATT comfortably for this task. It allows the most flexible distribution of test activities from a central test repository that allows you to address various test systems, depending on which system is currently active.

These CATT functions can support rollouts of R/3 implementation projects for multinational enterprises that operate with various R/3 releases.

14.4.14   **SAP Business Workflow and CATT**

The most different functions, being necessary for tests in the R/3-environment, are supported by CATT. In connection with the SAP Business Workflow, the option is given, to process recorded individual function procedures (Customer master record processing…). In particular this is useful for processes, in which very often transaction procedures are to be processed in the same way. By CATT the contents to be maintained can then be coordinated [Berthold99/341].

All usage fields of CATT within the SD-implementation described above, are in detail depicted in the book "Testing SAP R/3-systems" edited 2000 by Addison-Wesley-Longmann. There, the reader comes to know the entire functionality of CATT with test scenarios. It is explained, how the CATT-usage shall be organized in an R/3-system, and shows, how CATT shall be applied to the single phase activities of the best known procedure modules mentioned above. Further more, the CATT-usage for the maintenance and further development of the R/3-system are there described in detail.

14.5   **Advantages of the usage of CATT with the implementation of the SAP R/3 module DS**

The procedures described produce numerous advantages, as summarized in the following:

- CATT as an integral element of the procedure model
- Simplification of release updates by transferring CATTs
- More rapid ability of the SAP R/3 System to contribute to the success of the enterprise.
- Fewer maintenance activities
- Ability to remain within cost and time constraints
- Greater ability to control consulting firms
- Support for knowledge transfer
- Reduced training efforts

14.6 Application example of CATT in the module SD

In this section, on the basis of the credit limit check in the module SD, the functionality of CATT in regard of the maintenance of customizing tables is explained. By means of the here stated application example, the user is enabled, to test customizing tables in the light their effects in the system and subsequently to renew their starting situation.

14.6.1 Maintaining customizing tables with screens or transactions

Creating a test module for the New Table Entries

1. Menu path: UTILITIES → ABAP WORKBENCH
2. Menu path: TEST → TEST WORKBENCH → CATT PROCEDURES
3. Record
4. Enter the following data in the appropriate fields:

Fields	Data
Transaction code	0VA8 (=automatic credit-limit check)

New entries

Enter the following data in the appropriate fields:

Fields	Data
KKBR	0001
RKI	001
KG	01
Season factor	50

1. Save

Cancel

Do you really want to cancel?: Yes.

Back

Recording ends

Enter

Enter the following data in the appropriate fields:

Fields	Data
Short text	Customizing profile: automatic credit limit check
Key word	A020020435
Application	SD
Subapplication	BF
Component	CM

Menu path: TEST PROCEDURE → CHECK

Save without checking

Creating a test module for existing Table Entries

1. Menu path: UTILITIES → ABAP WORKBENCH
2. Menu path: TEST → TEST WORKBENCH → CATT PROCEDURES
3. Record
4. Enter the following data in the appropriate fields:

Fields	Data
Transaction code	0VA8 (=automatic credit-limit check)

5. Itemization
6. Enter the following data in the appropriate fields:

Fields	Data
KKBR	0001
RKI	001
KG	01

7. Continue with enter

8. Detail
9. Enter the following data in the appropriate fields:

Fields	Data
Season factor	60

10. Save
11. Back
12. Back
13. Recording ends with enter
14. Enter the following data in the appropriate fields:

Fields	Data
Short text	Customizing profile: credit limit check
	Automatic
Key word	A020020435
Application	SD
Sub application	BF
Component	CM

15. Menu path: TEST PROCEDURE → CHECK
16. Save without checking (F11)

Setting a test procedure from both Test Modules

1. Menu path: UTILITIES → ABAP WORKBENCH
2. Menu path: TEST → TEST WORKBENCH → CATT PROCEDURES
3. Copy
4. Enter the following data in the appropriate fields:

Fields	Data
Test procedure	KBC00036

5. Select *copy*

6. Enter the following data in the appropriate fields:

Fields	Data
Short text	Customizing profile: automatic credit limit check
Key word	A020020435
Application	SD
Subapplication	BF
Component	CM

7. Select *functions*

Note: Definition of the key fields as import parameters

8. Enter the following data in the appropriate fields:

Fields	Data
Function	CHETAB
Object	T691F

9. Enter the following data in the appropriate fields:

Fields	Data
Object T691F	Double-click

10. Enter the following data in the appropriate fields:

Fields	Data
Credit control area	&

11. Press the enter key

12. Enter the following data in the appropriate fields:

Fields	Data
Parameter name	&KKBER
Suggested value	0001

13. Continue

14. Enter the following data in the appropriate fields:

Fields	Data
Credit management: risk class	&

15. Enter

16. Enter the following data in the appropriate fields:

Fields	Data
Parameter name	&CTLPC_CM
Suggested value	001

17. Continue

18. Enter the following data in the appropriate fields:

Fields	Data
Credit group process	&

19. Enter

20. Enter the following data in the appropriate fields:

Fields	Data
Parameter name	&CRMGR_CM
Suggested value	01

21. Back

22. Enter the following data in the appropriate fields:

Fields	Data
Reference 1	KBC00249

23. Menu path: EDIT → BREAK DOWN REFERENCE
24. Enter the following data in the appropriate fields:

Fields	Data
Reference 2	KBC00250

25. Menu path: EDIT → BREAK DOWN REFERENCE
26. Double-click on the TCD line, and then replace all key parameters with an ampersand (&).
27. Save

Creating a test module

1. 1. Menu path: UILITIES → ABAP WORKBENCH
2. 2. Menu path: TEST → TEST WORKBENCH → CATT PROCEDURES
3. 3. Create
4. 4. Fill the following fields in the attribute maintenance screen:

Fields	Data
Short text	Customizing profile: automatic credit limit check
Key word	A020020435
Application	SD
Sub application	BF
Component	CM

5. Now you must navigate to the function screen, where you define the key fields of the table as import parameters.

6. Enter the following data in the appropriate fields:

Fields	Data
Function	SETTAB
Object	T691F

7. Confirm with enter

8. Double-click on object *T691F*

9. Enter the following data in the appropriate fields:

Fields	Data
Credit control area	&

10. Confirm with enter

11. Enter the following data in the appropriate fields:

Fields	Data
Parameter name	&KKBER
Suggested value	0001

12. Continue

13. Enter the following data in the appropriate fields:

Fields	Data
Credit management: risk class	&

14. Enter

15. Enter the following data in the appropriate fields:

Fields	Data
Parameter name	&CTLPC_CM
Suggested value	001

16. Continue

17. Enter the following data in the appropriate fields:

Fields	Data
Credit group process	&

18. Enter

19. Enter the following data in the appropriate fields:

Fields	Data
Parameter name	&CRMGR_CM
Suggested value	01

20. Continue

21. Display all fields

Note: You can fill all the fields displayed here with any values you wish. If the Customizing table already contains values, enter a different one overwrites the previous value. If the *field contents* field remains blank, the entry in the Customizing table is not modified.

22. Enter the following data in the appropriate fields:

Fields	Data
Season factor in %	50

23. Back

Note: To ensure that the SETTAB setting are not reset, you must enter the RESTAB% function. Doing so deletes the resetting data for the SETTAB settings.

24. Enter the following data in the appropriate fields:

Fields	Data
Function	RESTAB
Object	%

25. Menu path: TEST PROCEDURE → CHECK

26. Save without checking

27. Local object

28. Back

Processing the test module

1. Menu path: UTILITIES → ABAP WORKBENCH

2. Menu path: TEST → TEST WORKBENCH → CATT PROCE-DURES

3. Enter the following data in the appropriate fields:

Fields	Data
Test procedure	KBC00251

4. Execute

5. Enter the following data in the appropriate fields:

Fields	Data
Log type: long	Check
Processing mode: foreground	Check
Variants	Without
Credit group process	02
Credit management: risk class	Blank
Credit control area	Blank

Note: If you leave fields for the import parameters blank, the system uses the suggested value. The example maintains the Customizing table for credit group process 02, risk class 001 (= suggested value) and credit control area 0001 (=suggested value).

6. Execute

7. Enter the following data in the appropriate fields:

Fields	Data
KBC00251	Check

8. Menu path: EDIT → EXPAND NODES

9. Back

Note: The test module undergoes a check when you call transaction 0VA8 (= view for maintenance of automatic credit limit control). There you should maintain a season factor of 50% for credit group 02.

14.7   Example of an operational test scenario in regard of an R/3-implementation with CATT

During the R/3 implementation project the project leadership decided to submit the system settings made during the detailing and implementation phase of the project to a larger integration test. The implementation project had already advanced significantly, so the project leadership had two basic options for executing a test. It first discussed having the persons responsible for the components execute the test. This option involved a suggested testing length of about five days. Project leadership regarded this period as too long, given the already tight schedule of the R/3 implementation project, and as too expensive. This clear temporal restriction moved the leaders of the project to decide for the use of a CASE tool, led by an experienced consultant. After examining the testing tools available, project leadership decided to use the Computer Aided Test Tool from SAP. The most important criteria for the decision? CATT was a tool developed by SAP itself, so it was already tailored to SAP software in its details and functions, and as part of the ABAP Development Workbench. The company did not need to buy a new software product. In a workshop, the project leaders decided on the testing goal: error-free execution of diverse integration tests based upon various test scenarios developed by the component teams.

Within the R/3-implementation project, the most different business processes have been graphically displayed among the usage of the mentioned modeling tool, for the preparation of the integration test. These business processes were subsequently divided in single SAP-transactions. Every single transaction was recorded by CATT. The created test components were integrated in corresponding, the desired process-orientated functionality displaying test procedures, in accordance with the defined business processes.

The module teams had defined the system in such a way, that the business processes, contained in the test cycles, could be continuously and in an integrative way processed. The adequate master data were maintained by the responsible employees.

Following modules were implemented in the scope of the implementation project:

- MM Materials Management
- EC External Accounting
- PP Production Planning
- SD Sales and distribution

The following describes the test cycle that integrates these four components as an example:

MM (Materials Management) Component

Materials Management first generates two orders for materials. The orders involve material no. 100190 and its packaging, material no. 10090.

The KUNDE Company (vendor no. 500141) provides material 100190; the WKW Company (vendor no. 500406) provides the packaging.

At the end of the appropriate transaction, the system automatically assigns unique order numbers to each order. With the order numbers, the user now wishes to post a goods receipt. The posting of the goods receipt results in another unique document number for each order.

EC (External Accounting) Component

Based upon the previously generated document numbers for each goods receipt, the EC component adds an invoice from each vendor. The R/3 System assigns an invoice number that permits unique identification of each vendor invoice. The External Accounting component posts the invoices.

PP (Production Planning) Component

The PP component then creates process orders requesting production of the end product, material no. 10019010090, for the client: ABC. The ABC Company exists in the system under customer no. 100263. Based upon the process order number assigned by the system, the PP component also issues a confirmation for the entire process order. The result of the confirmation transactions should save the process order and ensure that all previous goods movements run without errors.

SD (Sales an Distribution) Component

The SD component generates a customer order for the ABC customer (customer no. 100263), with an order for end product 10019010090. The system stores the customer order as a standard order number. Based upon the standard order number, the SD component creates a delivery with automatic batch determination for the ABC Company. The result of this transaction stores the previously created delivery in the system, a delivery identified by a unique delivery number. Finally, the SD component creates an invoice for the delivery, and the R/3 system assigns a unique document number to the invoice. The document number should remain available, in case you want to expand the test process.

14.7.1   Recording test modules

This step maps all the transactions listed above in CATT with the data provided by the component experts. A test module is created for each individual transaction. The following list of test modules results:

- CATT test module Create purchase order
- CATT test module Post goods receipt for purchase order
- CATT test module Enter invoice
- CATT test module Create process order
- CATT test module Create process order confirmation
- CATT test module Overhead calculation
- CATT test module WIP calculation
- CATT test module Variance calculation for production orders
- CATT test module Actual settlement order
- CATT test module Create sales order
- CATT test module Create delivery
- CATT test module Create billing document

14.7.2   **Executing the transaction tests**

Here you start the various test modules individually and submit each to CATT testing. The tests change the data and produce results. If the module test uncovers errors in the test modules (the transactions), the data generated by the transactions, or the results of the transactions, you must correct the errors. As noted above, CATT offers you the option of isolating the error and correcting it once you have identified it. Only when no difference exists between the expected result of a transaction and the actual result, can you consider the test module as correct. At that point you can set the test status to *released* in the attribute maintenance screen.

Before you can combine the individually generated and checked test modules into a test procedure, you must ask yourself how to provide them with parameters. Doing so allows the test modules to exchange data with each other during the test procedure and makes them as widely useable and flexible as possible. Flexibility means that you can use the modules in other procedures.

14.7.3   **Generating the test procedure**

For the creation of the test procedures the different test components must be referenced. The test procedure displays a complete operational or administrative process with all prerequisites, which are necessary for a certain test. "Create standard order" requires e.g. as prerequisites "*Create customer*", "*Create material*" as well as "*Determine conditions*".

A test procedure can integrate any test modules to a total application or a specific functionality of an application.

14.7.4   **Executing the test procedure**

Now you can start the test procedure you have created. If errors occur in the procedure after it starts, repeat the test until you can identify and correct the errors. The integration test has run successfully when the test procedure completes with no errors and meets the goals of the testing design.

14.8   Management of CATT test cases

The execution of a test is an extremely important activity within an R/3 implementation project. It helps to reduce the risk that errors might reduce benefits to the business. Software testing seeks to find such errors. An error exists when execution of a test uncovers an expected result that does not match the actual result. A test process is successful either when it discovers existing errors or shows the absence of errors in the software. SAP offers two tools to support testing: the Computer Aided Test Tool (described above) for generating, maintaining, automatically executing, and logging test procedures, and the Test Workbench. SAP delivers both CATT and the Test Workbench with the R/3 System. SAP developed the Test Workbench to increase the error assessment rate. The tool plans, controls, and documents test processes.. You can use the Test Workbench to organize and manage automatic tests, manual tests, and external applications created with CATT. The process outlines the test scenarios into test catalogues, test plans, and test packages.

You can also link documentation on the test to the Test Workbench. The test administrator plans the entire testing process and creates a *test catalogue*. Beginning with the test catalogue, the test administrator subdivides the tasks further, to the level of test packages, which the administrator assigns to the persons who execute the tests.. According to such a plan, persons other than the test administrator should also perform testing.

With all its functions, the Test Workbench enables comfortable maintenance of the various test releases beyond the current release level of the R/3 System. The Test Workbench fulfills the following tasks:

- Clear specification of testing standards

- Complete documentation of test execution

- Enabling of test evaluation

- Guaranteed execution of subsequent tests

Experience has shown that the use of this tool is helpful even with a small number of test procedures. Even with a small number, the Test Workbench enables you to create a clear structure for all test procedures. You must hold to the following order for processing:

Step 1:   Create the test catalogue

Step 2:   Determine the structure of the test catalogue

Step 3:   Link CATT test procedures to the test catalogue

Step 4:   Create the test plan

Step 5:   Create the test packages

Step 6:   Assign test packages to the testers

Step 7:   Execute the test packages

Step 8:   Perform status analysis after the test

The described usage fields of the Test Work Bench are also explained in detail in the book "Testing SAP R/3-systems", by Gerhard Oberniedermaier and Marcus Geiss, ISBN 0-201-67517-X, edited 2000 in the "SAP-Series" by Addison-Wesley.

This book is a detailed practical guide for all R/3 consultants und ABAP programmers to the use of CATT, SAP's computer aided test tool, during the implementation of SAP R/3. It tells how to automate tests and optimize R/3 business processes, use CATT tools at each stage of an R/3 implementation, and how to use CATT tools in tasks such as testing transactions, testing database field values, configuring system tables, and generating test and training data.

15

Practice case study 1: Commission procedure

In this practice orientated case study it is illustrated, how all invoices included in the SAP R/3-system, with the corresponding commission parts, can be examined and automatically transmitted in a commission requirement. In the following, a practice proved development of a new transaction called ZPRO is represented, which controls a report processing the examinations as well as the creation of commission requirements by Call Transaction.

15.1 Procedure description of commissions

Fundamentally, commission requirements developed out of project day business of a company, are in practice generated by batch jobs in the background. A cyclic job must thereto be established as first step, which independently processes e.g. at the end of month (o1.o2 00:00) and records all documents being up to this time cleared in the SAP system and creates on their basis the corresponding commission orders. Documents, whose accounting document includes the state "cleared", are determined as automated commission stock and automatically transferred to a commission requirement.

The transaction ZPRO to be programmed serves thereby only for the display of the commission stock and not for its creation. All commissions (cleared, not cleared, confidential) can be displayed. For every commission recipient variants are to be generated in the ABAP/4-variant catalog, which then create the limitations required.

For example:

Variant 1: to the commission taker xy

Variant 2: to the commission taker yz

After the creation of the commission requirements of the SAP R/3-system, the specific stock is displayed by transaction VF04 (invoice stock order referred) and by stating the corresponding commission recipient. Through a simulation, to be correspondingly programmed, a list of invoices or but of the deficient orders to be corrected, can be generated ahead of the edition of a regular invoice takes place.

15.2

Report ZVPROV01 – Transaction (ZPRO)

The transaction ZPRO serves merely for the display of the commission stock, no commission orders are established in dialogue. In this report the document flow to all accounting-and invoice documents is inspected. With the parameter "Directly posting" all determined commissions without dialogue are automatically transferred to a commission requirement. This is nevertheless only possible in the batch.

By means of commission receiver, invoice date (from...to) and the posting period (month, year) the expected commissions can be selected. The invoice types credit/debit memo are adapted as standard proposal by the program and can, if required, be expanded or change in the dialogue. Though they should not be deleted, since the selection determines the following order type.

The invoice types for the debit memo receive the order type ZLPA in the commission requirement. (Credit memos receive the order type ZGPA in the commission requirement.)

The established list is graded referring the currencies to the commission recipient. All positions of the invoice including the condition type ZPRN (commission) are represented like that. The totals-lines are summarized referring currencies.

15.3

Program description of the necessary user exits

During the creation of the commission requirement the commission receiver (ZP) is copied to all partner roles (AG/WE/RE/RG). As prerequisite it is to be aware of that the customers must be maintained as debtors.

15.3.1   User Exit: Take over of partner roles of the invoice of the commission recipient

Program SAPMV45A

INCLUDE MV45AFZB.

```
****************************************************************
*# Take-over of the ordering party out of the commission
* invoice receipt
****************************************************************
FORM USEREXIT_MOVE_FIELD_TO_TVCOM_H.
* Examples:
* TVCOM-zzfield = VBAK-zzfield2.
* TVCOM-zzfield = VBKD-zzfield2.
*# Take-over of the commission receipt of invoice
   PERFORM ZZ_GET_PROV_KUNDE
*# end of insertion
ENDFORM.
*&--------------------------------------------------------*
*&      Form    ZZ_GET_PROV_KUNDE
*&--------------------------------------------------------*
*# Take-over of the commission receipt of invoice
*--------------------------------------------------------*
FORM ZZ_GET_PROV_KUNDE.
*
DATA: PROVNR LIKE XVBPA-LIFNR. " Number commission receiver
*
*# only with invoice take-over for commission requirement
   CHECK: VBAK-AUART = 'ZGPA' OR VBAK-AUART = 'ZLPA',
          VBAK-KALSM = 'ZVAA05'.
*# Read commission receiver
   READ TABLE XVBPA  WITH KEY PARVW = 'ZP'.
```

```
        CHECK SY-SUBRC = 0.

        PROVNR = XVBPA-LIFNR.
*
* Test, whether debtor exist sob
* (ext. program on grounds of table definition.)
        PERFORM ZZ_CHECK_DEBITOR(ZVPROV01) USING PROVNR.

        CHECK SY-SUBRC = 0.
*

LOOP AT XVBPA.

      CHECK: XVBPA-PARVW = 'AG' OR

             XVBPA-PARVW = 'RE' OR

             XVBPA-PARVW = 'RG' OR

             XVBPA-PARVW = 'WE'.

      XVBPA-KUNNR = PROVNR.

      MODIFY XVBPA.

    ENDLOOP.

ENDFORM.                        " ZZ_GET_PROV_KUNDE

*eject.
```

15.3.2   User Exit: Copy control invoice on commission requirement

***INCLUDE RV61AFZA.

In this User Exit the invoice types for the calculation scheme ZVAA01 are taken over from the calc.scheme ZVAA01. The in.type PR00 changes to ZPRG (ZPRL) and the In.type ZPRN turns to ZPRG (ZPRL).

```
*-------------------------------------------------------------

*       FORM USEREXIT_PRICING_COPY

*-------------------------------------------------------------*

*       modify KONV before copying

*-------------------------------------------------------------*
```

FORM USEREXIT_PRICING_COPY.

```
* the following example modifies the calculation

* rule within invoices

* for condition type HD00

.....

*# Determine conditions to the commission procedure

PERFORM ZZ_CHECK_AND_GET_PROVISION.

ENDFORM.

*&------------------------------------------------------------*

*&      Form   ZZ_CHECK_AND_CET_PROVISION

*&------------------------------------------------------------*

*   Determine condition for the commission procedure

*-------------------------------------------------------------

FORM ZZ_CHECK_AND_GET_PROVISION.

*

* tables declaration

TABLES: VBRK, VBRP.

*

* data declaration

DATA: PROV_BAS LIKE KONV-KBETR,
```

```
                          PROV_WRT LIKE KONV-KWERT,          " TYPE P DECIMALS 6,

                          PROV_ERG TYPE P DECIMALS 2.
* Check on commission procedure
  CHECK: VBTYP_NEW = 'K' .
* newly define key for the condition , so that it can be
* taken over.
  CASE    KONV-KSCHL.

    WHEN 'PR00'.

      KONV-KSCHL = 'ZPRP'.

      PROV_WRT = KONV-KWERT.       " Com value
* determine number of the invoice document
      SELECT * FROM VBRK WHERE KNUMV = KONV-KNUMV.

        EXIT.

      ENDSELECT.
* Position of the invoice
      SELECT SINGLE * FROM VBRP WHERE VBELN = VBRK-VBELN

                                    AND    POSNR = KONV-KPOSN.

      CHECK SY-SUBRC = 0.
* Determine com. basis  BONBA / Com.value * amount
      PROV_BAS = ( VBRP-BONBA * 10000 ) / PROV_WRT.

      KONV-KBETR = ( PROV_BAS *  KONV-KBETR ) / 10000.

      KONV-KSTAT = 'X'.

    WHEN 'ZPRN'.

      KONV-KSCHL = 'ZPRG'.

      KONV-KBETR = KONV-KBETR * -1.

      CLEAR KONV-KSTAT.

  ENDCASE.

ENDFORM.                 " ZZ_CHECK_AND_GET_PROVISION

....##
```

15.3.3 User Exit: Selection conditions and procedure for the determination of the commission stock

Procedure logic - Report ZVPROV01

```
* provide invoice document header data transmit to
* internal tale
   SELECT * FROM BKPF INTO I_BKPF WHERE BELNR IN BELNR
                     AND BLART    = 'RV'
                     AND MONAT    IN MONAT
                     AND GJAHR    IN GJAHR.
```

*   Collect only documents with commission

```
* ...only cleared invoices
    SELECT * FROM BSEG WHERE BELNR = I_BKPF-BELNR
                     AND    AUGBL NE ''.
* ...take corresponding invoice
    SELECT SINGLE * FROM VBRK WHERE VBELN =  P_FAKNR
                     AND    FKDAT IN FKDAT
                     AND ( FKART IN FKALAST
                     OR     FKART IN FKAGUT ).
*
* ...provide corresponding order
   PERFORM GET_VBFA_AUFNR USING P_FAKNR
                     CHANGING AUFNR.
   SY-SUBRC - 1.         " if no order -> End negative
*
* ...process only documents with commission recipient
   SELECT SINGLE * FROM VBPA WHERE VBELN = AUFNR
                AND    POSNR   = '000000' " Position number
                AND    PARVW   = 'ZP'.    " Partner roles
   IF SY-SUBRC NE 0.
*
* ...examine document flow for the order credit memo
```

```
* requirement established??
    LOOP AT I_VBFA WHERE VBELV  = AUFNR
                       AND ( VBTYP_N = 'K' OR
                             VBTYP_N = 'L' ).
      IF AUF_CNT > 2 AND GUTFLG = 'X'. EXIT. ENDIF.
      PERFORM CHECK_VBFA_PROVNR USING I_VBFA-VBELN P_FAKNR.
*

*...credit memo requirement must not yet be established
      SELECT SINGLE * FROM VBAK WHERE VBELN = I_VBFA-VBELN
                      AND ( AUART = PROV_ZGPA       " 'ZGPA'.
                      OR    AUART = PROV_ZLPA ).    " 'ZLPA'.
      CHECK SY-SUBRC = 0 AND AUF_CNT > 0.
* .. if o.k.
      CHECK SY-SUBRC = 0.
*

* .. secure all in internal tables
      PERFORM SAVE_DATA_EXTRACT.
```

15.4 Expansion of the copy conditions

Expansion of the copy conditions to ZVAA05, this routine is used within the copy control INVOICE → ORDER for the order types ZGPA and ZLPA.

```
*------------------------------------------------------------
*       FORM DATEN_KOPIEREN_956
*------------------------------------------------------------
*
* Order header of invoice
* Following work areas are available:
*
* VBAK - header of the order to be expanded
* CVBRK - Model header
*
```

```
*------------------------------------------------------------*
FORM DATEN_KOPIEREN_956.
LOCAL: VBAK-VBTYP,
       VBAK-VBELN,
*      vbak-kalsm,
       VBAK-ERDAT,
       VBAK-ERZET,
       VBAK-ERNAM.
DATA:  DA_VBAK_STAFO LIKE VBAK-STAFO.
*
  IF CVBRK-VBELN IS INITIAL.
    MESSAGE A247 WITH '956'.
  ENDIF.
  DA_VBAK_STAFO = VBAK-STAFO.
  MOVE-CORRESPONDING CVBRK TO VBAK.
  VBAK-STAFO = DA_VBAK_STAFO.
* Only this dispatch condition filed in the
* customizing counts
  IF NOT TVAK-VSBED IS INITIAL.
    VBAK-VSBED = TVAK-VSBED.
  ENDIF.
*
  VBAK-KALSM - 'ZVAA05'.
* equalize header fields split criteria
  PERFORM ZZ_PROVISIONS_SPLIT.
ENDFORM.
*eject
*&-----------------------------------------------------------*
*&      Form   ZZ_PROVISIONS_SPLIT
*&-----------------------------------------------------------*
* equalize header fields split criteria
```

```
*-------------------------------------------------------*

FORM ZZ_PROVISIONS_SPLIT.

TABLES: KNVV, KNA1.

*

* delete not relevant data CVBRK

  CLEAR CVBRK-ZUKRI.              " Connection criteria

  CLEAR CVBRK-FKDAT.             " Invoice date f. index.

  CVBRK-LANDTX = 'DE'.           "# Test

*

* delete not relevant data VBAK

  CLEAR VBAK-XBLNR.      " delete reference document number

  CLEAR VBAK-VSBED.      " Delete dispatch conditions

  VBAK-KALSM_CH = 'ZVAA05'. " Allocate calc. scheme charge

*

* determine commission recipient.

  SELECT SINGLE * FROM VBPA WHERE VBELN = VBAK-VBELN

                            AND    POSNR = '000000'

                            AND    PARVW = 'ZP'.

  CHECK SY-SUBRC = 0

  VBAK-KNKLI = VBAK-KUNNR = VBPA-LIFNR.

                            " Allocate commission recipient

  CVBRK-KUNAG = VBPA-LIFNR. " Allocate commission recipient

  CVBRK-KUNRG = VBPA-LIFNR. " Allocate com. recipient

  CVBRK-KNKLI = VBPA-LIFNR. " Allocate com. recipient

*

* Determine payment conditions

  SELECT * FROM KNVV WHERE KUNNR = VBPA-LIFNR

*                    AND    VKORG = VBAK-VKORG

                     AND    VTWEG = VBAK-VTWEG

                     AND    SPART = VBAK-SPART.

    VBAK-VKORG = KNVV-VKORG.
```

```
        VBAK-VTWEG = KNVV-VTWEG.

        VBAK-SPART = KNVV-SPART.

        CVBRK-VKORG = KNVV-VKORG.

        CVBRK-VTWEG = KNVV-VTWEG.

        CVBRK-SPART = KNVV-SPART.

        CVBRK-ZTERM = KNVV-ZTERM.

        CVBRK-INCO1 = KNVV-INCO1.     " IncoTerms 1 delete

        CVBRK-INCO2 = KNVV-INCO2.     " IncoTerms 2 delete

      EXIT.

    ENDSELECT.

  ENDFORM.                           " ZZ_PROVISIONS_SPLIT
```

16

Practice case study 2: Revenue account finding

16.1   Survey

In the following menu items, the control of the revenue account finding is defined for the transmission of invoice values from the module SD to the module FI. The revenue account finding is processed with the help of the condition technique. Following criteria are of importance for the control of the revenue account finding:

- Account finding table: In it the criteria are determined, upon which the account finding shall depend.

- Access sequence: Here, it is determined, in what sequence the fields of the invoice document (header and position) are to be read.

- Account finding type: The account finding type is a component of the condition technique. It must be at minimum one account finding type be established and allocated to one account finding scheme.

- Account finding scheme: This scheme contains the required account finding types, with which the SAP-system can automatically find the revenue accounts.

16.2   Check of the account assignment relevant master data

The account finding can be referred to following master data fields:

- Account group material in the material master record

- Account group customer in the customer master record

Suggestive divisions for materials can be for example:

- Revenues for services (material type DIEN)

- Revenue for packages (material type VERP)

- Revenue for products (material type FERT)

- Revenues for trading goods (material type HAWA)

Suggestive divisions for customers can be:

- Revenues for external customers inland

- Revenues for external customers foreign countries

- Revenues for connected enterprises (group companies)

- Revenues for customers of EU-member states

- Revenues for customers of EFTA-states

Note: The revenue accounts are divided in financial accounting (System FI), in order to receive certain evaluations and /or to coordinate to the financial statement (System CO). Therefore, the account finding respectively the account groups are to be determined in coordination with the financial accounting and cost accounting.

Activities

In the first step, you branch into the revenue account finding in the IMG.

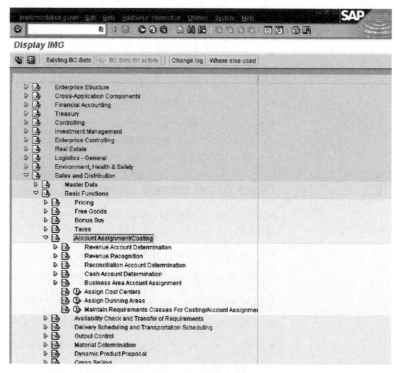

Illustration 16.1: IMG Revenue account finding-Picture 1 © SAP AG

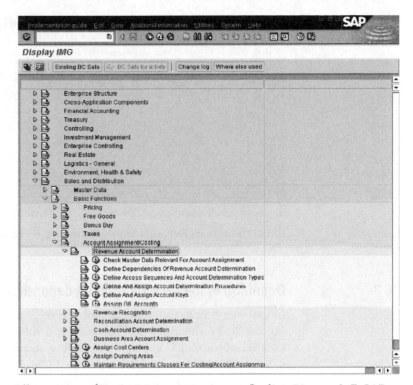

Illustration 16.2: IMG-Revenue account finding-Picture 2 © SAP
AG

1. Determine, which account groups for materials and cus-
    tomers you require, in order to arrange the master records
    for the financial statement.

2. Give a particular alphanumerical key, with up to two posi-
    tions, to the account group Material, and a name.

3. Give a particular alphanumerical key, with up to two posi-
    tions, to the account group Customer.

4. Assure yourself, that the account group keys are registered
    in the material master record and the customer master re-
    cord.

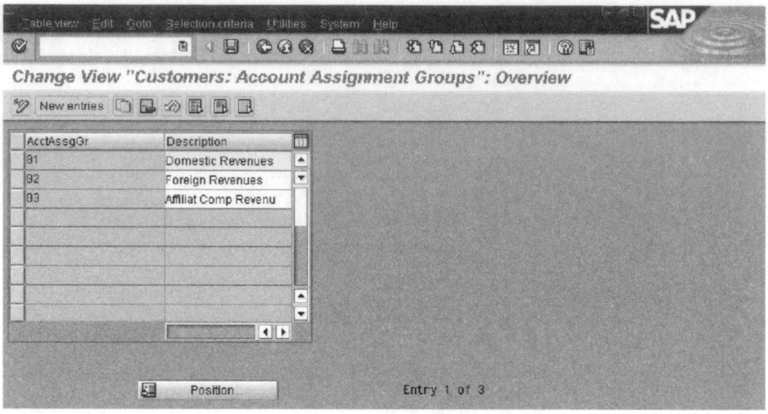

Illustration 16.3: Revenue account finding – change of Customer
account groups © SAP AG

16.3   Definition of the revenue account dependencies

In this menu item the dependencies of the revenue account finding are to be determined. Therefore you file the combination of
criteria, upon which the account finding should depend, in a account finding table.

You can select following criteria for the account finding table:

- Condition type
- Account plan
- Account group Customer
- Account group Material
- Account key
- Sales organization
- Sales path
- Plant

Activities

1. Check, in how far you can apply the applications included in the SAP-standard delivery in regard of the account finding table. Therefore you can let the existing account finding tables be displayed.

2. Define the account finding table: Consider during the creation, that you have to choose a key between 501 and 999 for the condition table.

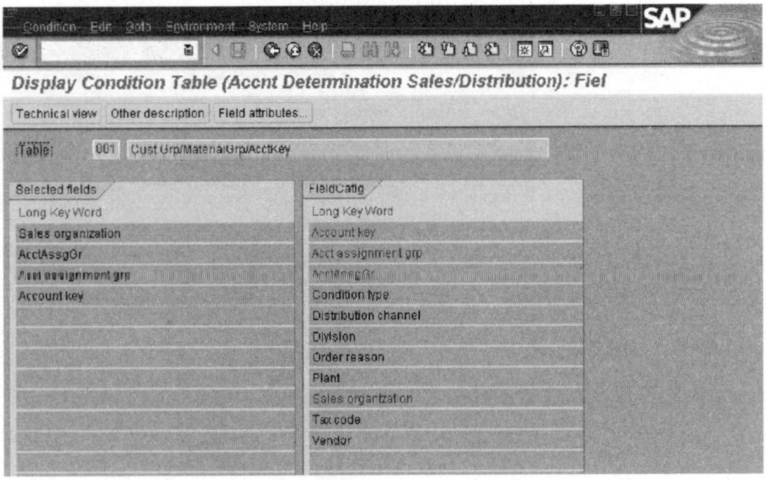

Illustration 16.4: Revenue account finding Condition table
© SAP AG

16.4   Definition of access sequences and account finding types

In this menu item you define the access sequences and the account finding types for the automatic revenue account finding. With the access sequence following is determined:

- The condition tables, with which the SAP-system shall gain access to the condition tables

- The sequence, with which the condition tables shall read

- The field contents, with which the condition types shall be read

In the account finding type you determine the control data access sequence and validity date.

Standard applications

In the SAP-standard application five access sequences are defined, as for example the access sequences:

- Account plan/Sales organization/Account group Customer/Account group Material/Account key

- Account plan/Sales organization/Account key

In the SAP-standard application a account finding type with the key "KOFI" is filed.

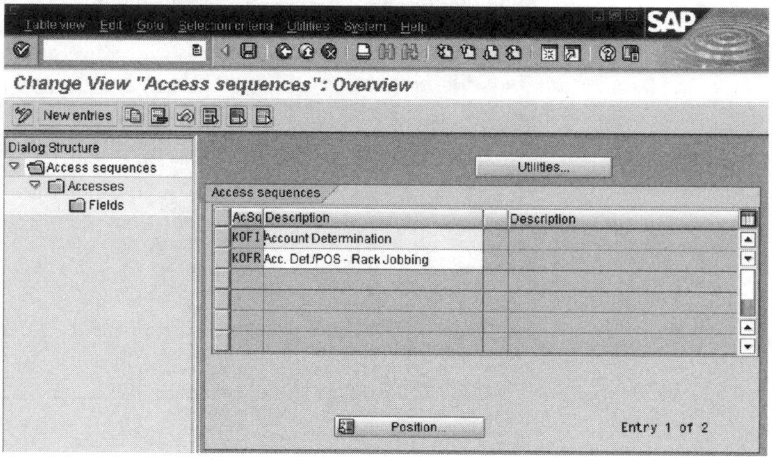

Illustration 16.5: Revenue account finding-Access sequences
©SAP AG

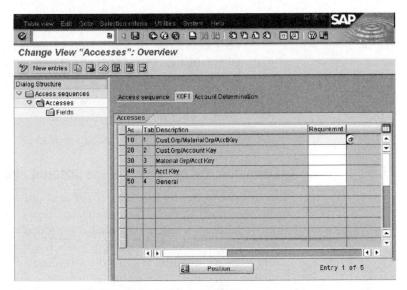

Illustration 16.6: Revenue account finding-Access to access sequence KOFI ©SAP AG

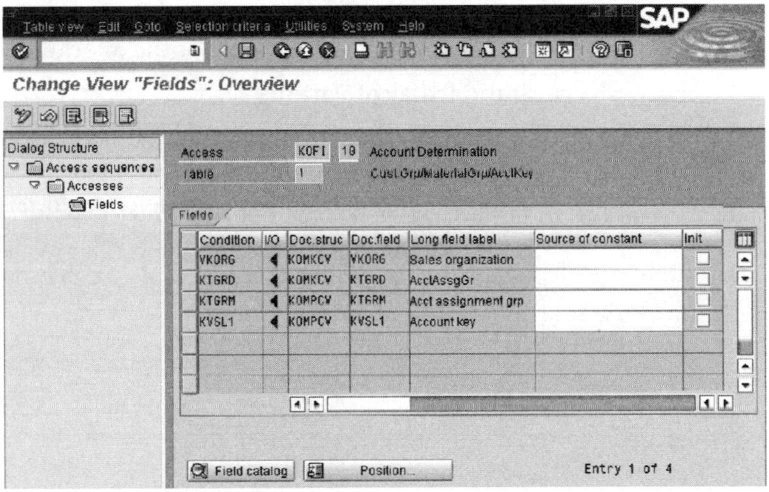

Illustration 16.7: Revenue account finding-Fields of the access KOFI ©SAP AG

With these items the SAP-system can already process the mechanical account finding.

Activities

1.   Check, in how far you can use the access sequences contained in the SAP-standard version, and the account finding types

2.   Define your access sequence. Select therefore the relevant criteria for a new access sequence

3.   Define your account finding types.

16.5   Definition and allocation of the account finding scheme

In this menu item you define your account finding schemes and allocate these to the different invoice types. In an account finding scheme you define, in which sequence the SAP-system shall read these account finding types, which are applied for the revenue account finding. The account finding schemes are allocated to the particular invoice types, for which a corresponding account finding shall be ensued.

Prerequisites

The invoice types must be defined. The defined invoice types are automatically proposed for the allocation.

Standard applications

In the SAP-standard application an account finding scheme is already defined with the key "KOFI00". With this scheme the automatic account finding can be covered.

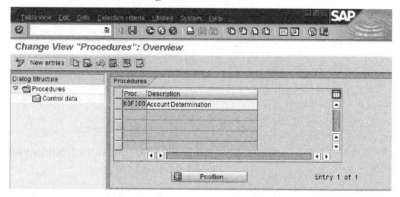

Illustration 16.8: Revenue account finding-Account finding scheme KOFI00 ©SAP AG

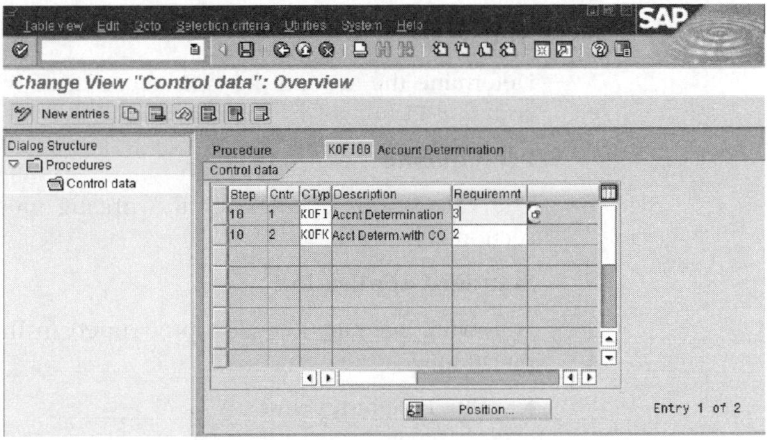

Illustration 16.9: Revenue account finding Control of the account finding scheme KOFI00 ©SAP AG

Activities

1. Check, in how far you can apply the applications to the account finding schemes included in the SAP-standard delivery

2. Define your account finding scheme and give the particular account finding types, relevant for this account finding

3. Allocate the account finding schemes to the invoice types. State hereby, if necessary, for cash payments the account key of the cash clearing.

16.6 Definition and allocation of the account key

In this menu item you define your account keys and allocate these to the condition types in the calculation schemes of the pricing. With the account keys you summarize equivalent accounts in the financial accounting. The SAP-system finds the G/L accounts looked for, with the help of the account key. You can at any time allocate a separate key to every condition type of the pricing within a calculation scheme, in order to realize a detailed revenue account finding.

You can, for example, by that means, allocate a freight condition to a freight revenue account respectively allocate a surcharge for the packaging costs to a corresponding account for packaging revenues.

Notice

Determine the account keys in accordance with the financial accounting FI.

Prerequisite

The calculation schemes for the pricing must be defined (see section "Pricing").

Standard application

Following account keys are predefined in the SAP-standard application:

- ERF Freight revenues
- ERL Revenues
- ERS Sales deduction
- EVV Cash clearing
- MWS Value added tax

Activities

1. Check, in how far the applications or the account key contained in the SAP-standard application are useful for you.

2. Define the account key, by stating an alphanumerical key, including up to three positions, and a text description.

3. Allocate the revenue account key to the condition types in the calculation schemes. Thereto you can call up a list as usage proof of the account key. In this list it is displayed for every account key, to which condition type an account key has been allocated in which calculation scheme.

16.7   Allocation of G/L accounts

In this menu item you allocate the G/L accounts for the revenue account finding. The allocation must be made for every access sequence, being defined before. Depending on the key combination, different criteria are important for a G/L account. With the key combination "Account group customer/account key" a G/L-account is given e.g. in dependence on the following criteria:

- Application (key for the application S/D)
- Account finding type

- Account plan (of the module FI)

- Sales organization

- Account group customer

- Account key

Prerequisites

In the scope of the condition technique, access sequences and account-finding types as well as account-finding schemes and condition-tables must be defined in SD. The same is necessary for the account plan and the G/L-accounts in the module FI.

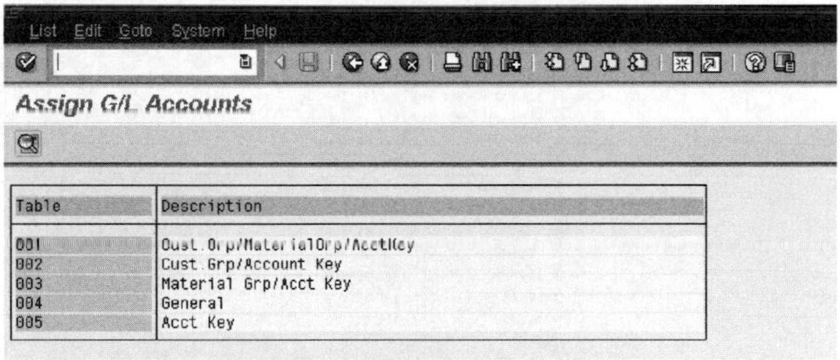

Illustration 16.10: Allocation of G/L-accounts -Picture 1 ©SAP AG

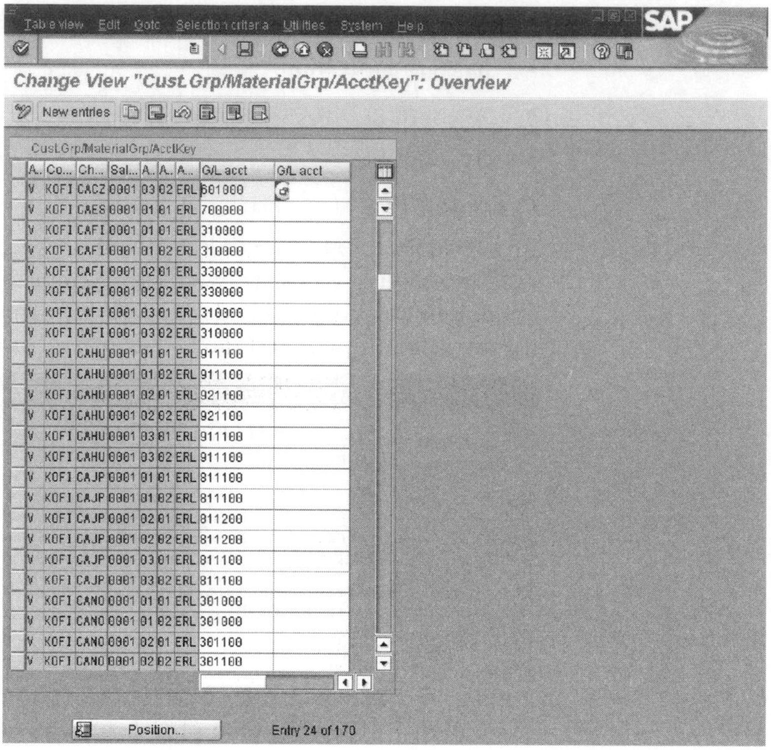

Illustration 16.11: Allocation of G/L-accounts – Picture 2 © SAP AG

Activities

1. Check, whether you can use the allocations included in the SAP-standard application

2. Allocate the G/L-accounts to the particular access sequence

Appendix A

R/3 Implementation: Classic Procedure Model

Phase 1: Organization and Design

Project preparation

- Initialize project
- Define enterprise goals for the use of R/3
- Perform initial entry of inventory into the R/3 System
- Learn the processes and function of the R/3 Systems
- Define business processes
- Compare functional requirements with the R/3 System
- Map the enterprise structure
- Define the goals and scope of the standardization process
- Set the implementation strategy
- Determine hardware requirements
- Set the project structure
- Set project standards and working methodology
- Set the system landscape
- Create a resource plan, schedule, and cost plan
- Approve the results of project preparation
- Create the project order
- Kick-off the implementation project

Set up the system landscape

- Set up systems and clients
- Set up user master data for project participants
- Set up control of clients and the Correction and Transport System
- Set up the system landscape

- Set up remote connections
- Change country-specific default settings
- Create the enterprise IMG
- Create Customizing projects

Train the project team

- Train the project team
- Learn R/3 functions

Set processes and functions

- Specify processes and functions based upon the reference model
- Set responsibilities for processes and functions
- Check input/output information objects
- Determine requirements of reporting
- Determine interfaces and system enhancements
- Set mapping of the enterprise structure
- Prototype selected processes and functions
- Specify the data processing design
- Agree on the conceptual and data processing design

Design interfaces and system enhancements

- Create a detailed description for the interfaces
- Create a detailed description for the system enhancements
- Create a plan for data transfer

Quality assurance and targeted design

- Check project structure and procedural organization
- Check adherence to project standards
- Check the system landscape
- Check the targeted design
- Check the descriptions of interfaces and system enhancements
- Check project planning

- Create checking log
- Trigger release of the next phase

Phase 2: Detailing and Implementation

Make global settings

- Learn global settings
- Adjust global settings

Map enterprise structure

- Check and adjust R/3 System organizational units

Map basic and master data

- Determine fields and contents for master data
- Execute system settings for master data
- Test system settings for master data
- Detail description of master data transfer

Map processes and functions

- Set fields and contents for processes and functions
- Execute system settings for processes and functions
- Test system settings for processes and functions
- Introduce selected processes and functions to the user departments
- Detail description of data transfer

Implement interfaces and system enhancements

- Create data transfer programs
- Implement interfaces
- Implement system enhancements
- Test data transfer program
- Test interfaces
- Test system enhancements

Map reporting system

- Determine information needs

- Check coverage of information needs
- Create solutions for open requirements
- Determine organization of the reporting system
- Test reporting system

Map archiving management

- Create a design for archiving management
- Perform system settings
- Test archiving transactions

Map authorization management

- Create authorization concept
- Implement authorization concept
- Test authorizations

Execute closing test

- Create testing design
- Create testing plan
- Perform testing activities
- Create report on closing test
- Perform a closing review with the user departments

Quality check of application system

- Check project structure and procedural organization
- Check adherence to project standards
- Check conversion of the target design
- Check implementation of interfaces and system enhancements
- Check reporting system
- Check archiving management
- Check authorization concept
- Check closing test
- Check project planning
- Create check log

Trigger release of the next phase

Phase 3: Preparation for Production

Prepare for productive operation

- Concretize configuration for the production system
- Obtain system fixtures and fittings
- Create user master records
- Create plans for data transfer

Develop user documentation

- Set structure, contents, and media
- Prepare for creation of user documentation
- Create a change design
- Create user documentation

Set up productive environment

- Install network
- Install hardware and software at workstations
- Install the R/3 System in the productive system

Train users

- Create training program
- Prepare training
- Perform training

Organize system administration

- Determine system administration
- Train system administrators
- Transfer data to the production system
- Transfer system settings and objects
- Perform data transfer
- Create manual data
- Trigger acceptance of data transfer

Quality check of the production system

- Check user documentation
- Check production environment
- Check performance of user training
- Check organization of system administration
- Check data transfer
- Check project planning
- Create a check log
- Trigger release of the next phase

Phase 4: Productive Operation

Support productive operation

- Support users at the start of production
- Establish permanent support by organizing a help desk

Optimize system use

- Monitor and optimize system use
- Make adjustments
- Close project officially

Project administration and project controlling

- Perform fine planning
- Determine project status
- Trigger correction measures
- Perform project reviews

System maintenance and release upgrades

- Perform system upgrade or release upgrade
- Perform release Customizing

Appendix B

Description of Tasks in Individual ASAP Phases

SAP developed the following ASAP procedure model for the implementation of R/3 software.

Phase 1: Project Preparation

1.1 Creating the Project Plan

1.1.1 Create the Project Order

1.1.1.1	Formulate the mission statement for the project
1.1.1.2	Determine the business drivers
1.1.1.3	Set the success identifiers for the business
1.1.1.4	Determine the success numbers for the project
1.1.1.5	Collate the contents of the project order
1.1.1.6	Approve the project order

1.1.2 Check and Detail the Implementation Strategy

1.1.2.1	Check the suggested implementation
1.1.2.2	Confirm the implementation method
1.1.2.3	Check the rollout strategy of the enterprise

1.1.3 Set Up the Work Environment of the Project Team

1.1.3.1	Plan the work environment
1.1.3.2	Set up the work environment

1.1.4 Determine the Project Structure

1.1.4.1	Set the organization and rolls
1.1.4.2	Assign employees to roles
1.1.4.3	Hold a hand-over meeting

1.1.5 Create the Project Plan

1.1.5.1	Create the project plan
1.1.5.2	Create the project budget plan

1.1.5.3	Create the project usage plan

1.1.6 Create the Training Plan for the Project Team

1.1.6.1	Check suggestions from the pre-project phase
1.1.6.2	Rework the training plan
1.1.6.3	Register the training team and set the training schedule

1.2 Project Procedures

1.2.1 Define Project Management Standards Procedures

1.2.1.1	Set the communications plan
1.2.1.2	Define project documentation
1.2.1.3	Create a plan for issue management
1.2.1.4	Create a plan for project-scope management
1.2.1.5	Set a plan for team-building activities
1.2.1.6	Define standards for project planning and controlling
1.2.1.7	Set the use of R/3 services
1.2.1.8	Determine quality assurance procedures

1.2.2 Define implementation standards and procedures

1.2.2.1	Develop standards for system configuration
1.2.2.2	Define a strategy for user training and documentation
1.2.2.3	Set the testing strategy
1.2.2.4	Define a strategy for service and support after implementation
1.2.2.5	Define system authorization standards for the project team
1.2.2.6	Set procedures for dealing with problems and errors
1.2.2.7	Define the approval procedure for system enhancements and modifications

1.2.3 Define the System Landscape

1.2.3.1	Determine the systems required

Phase 2: Business Blueprint

2.1 Project Management Business Blueprint

2.1.1 Agreement on the Project Status

2.1.1.1	Prepare status meetings
2.1.1.2	Hold status meetings
2.1.1.3	Check completion of assigned tasks
2.1.1.4	Correct project deviations
2.1.1.5	Adjust project plan

2.1.2 Steering Committee

2.1.2.1	Prepare steering committee meetings
2.1.2.2	Hold steering committee meetings
2.1.2.3	Check completion of assigned tasks

2.1.3 General Project Management

2.1.3.1	Change management
2.1.3.2	Hold team-building activities
2.1.3.3	Define user roles and responsibilities

2.2 Training the Project Team Business Blueprint

2.2.1 Hold Project team training

2.2.1.1	Work on training program in detail
2.2.1.2	Prepare training
2.2.1.3	Hold project team training
2.2.1.4	Check and evaluate knowledge gained from training

2.3 Develop the System Environment

2.3.1 Draft the Technical Design

2.3.1.1	Document system infrastructure and distribution
2.3.1.2	Define and document the print infrastructure
2.3.1.3	Document the network topology
2.3.1.4	Document the interface topology

2.3.1.5	Define change request management
2.3.1.6	Define a release management strategy
2.3.1.7	Define a desktop management strategy
2.3.1.8	Agree on the technical design

2.3.2 Set Up the Development Environment

2.3.2.1	Install basic hardware
2.3.2.2	Check the system environment
2.3.2.3	Set up and configure development client
2.3.2.4	Install desktop components for the project team
2.3.2.5	Set up user master records for the project team
2.3.2.6	Secure the development system
2.3.2.7	Install and configure printing services for the project team
2.3.2.8	Configure remote network connections
2.3.2.9	Set up remote connections to SAP

2.3.3 Set System Landscape

| 2.3.3.1 | Set up and configure the development client |
| 2.3.3.2 | Configure and test the transport system |

2.3.4 System Administration

2.3.4.1	Hold a Basis and systems workshop
2.3.4.2	Define system administration for the development system
2.3.4.3	Configure CCMS
2.3.4.4	Define a backup strategy
2.3.4.5	Check system administration functions
2.3.4.6	Set periodic procedures in system administration

2.3.5 Initialize the Implementation Guide (IMG)

| 2.3.5.1 | Create the enterprise IMG |
| 2.3.5.2 | Create an IMG project |

2.4 Organizational Structure

2.4.1 Determine the Enterprise Structure

2.4.1.1   Plan a workshop on the enterprise structure

2.4.1.2   Distribute recommendations on the enterprise structure

2.4.1.3   Hold a workshop on the enterprise structure

2.4.1.4   Recommend and agree upon the enterprise structure

2.5 Define Business Processes

2.5.1 Prepare Business Process Workshops

2.5.1.1   Plan business process workshops

2.5.1.2   Hold a workshop on the procedure for defining processes

2.5.2 Hold a Workshop on Global Requirements

2.5.2.1   Set global parameters

2.5.2.2   Set enterprise-specific standards

2.5.3 Hold Business Process Workshops

2.5.3.1   Determine requirements for business processes

2.5.3.2   Determine the need for enhanced R/3 functions

2.5.3.3   Determine the needs of the reporting system

2.5.3.4   Determine the interfaces required

2.5.3.5   Determine the requirements for database transfer

2.5.3.6   Determine required enhancements

2.5.3.7   Identify areas of low coverage

2.5.3.8   Detail descriptions and models of business processes

2.5.3.9   Determine the need for additional, detailed workshops

2.5.3.10   Plan detailed requirements workshops

2.5.4 Hold Detailed Requirements Workshop

    2.5.4.1   Determine detailed requirements

    2.5.4.2   Optimize business process definition and models

2.5.5 Complete the Business Blueprint

    2.5.5.1   Detail the project organization and rolls

    2.5.5.2   Collate blueprint documents

    2.5.5.3   Determine the baseline scope

    2.5.5.4   Examine the completeness of the blueprints

2.5.6 Check and Release the Business Blueprint

    2.5.6.1   Prepare blueprint presentation

    2.5.6.2   Execute the check and release

2.5.7 Plan User Training

    2.5.7.1   Set the scope, contents, and process of user training

    2.5.7.2   Work out the user-training plan

    2.5.7.3   Agree on the user-training plan

2.6 Quality Inspection: Business Blueprint

2.6.1 Perform Quality Inspection and Acceptance

    2.6.1.1   Perform a quality inspection

    2.6.1.2   Release the business blueprint

Phase 3: Implementation

3.1 Implement Project Management

3.1.1 Agree on Project Status

    3.1.1.1   Prepare status meetings

    3.1.1.2   Hold status meetings

    3.1.1.3   Check completion of assigned tasks

    3.1.1.4   Correct project deviations

    3.1.1.5   Adjust project plan

3.1.2 Steering Committee

3.1.2.1	Prepare steering committee meetings
3.1.2.2	Hold steering committee meetings
3.1.2.3	Check completion of assigned tasks

3.1.3 Output Planning for Production Support and Cut-Over

3.1.3.1	Fix a plan for production support
3.1.3.2	Fix a plan for cut-over

3.1.4 Project Management in General

3.1.4.1	Change management
3.1.4.2	Perform team-building activities

3.2 Training for the Project Team

3.2.1 Hold Project Team Training

3.2.1.1	Work out the training program in detail
3.2.1.2	Prepare training
3.2.1.3	Hold project team training
3.2.1.4	Check and evaluate knowledge gained from training

3.3 Baseline Configuration and Acceptance

3.3.1 Create Plans for Baseline Configuration

3.3.1.1	Refine baseline scope
3.3.1.2	Create configuration plan for baseline
3.3.1.3	Determine test cases
3.3.1.4	Create test plan for baseline
3.3.1.5	Assign resources
3.3.1.6	Agree on baseline configuration
3.3.1.7	Detail project IMG

3.3.2 Configure Global Settings and Enterprise Structure

3.3.2.1	Set up global settings
3.3.2.2	Set up organizational structure

| 3.3.2.3 | Include pre-defined settings |

3.3.3 Configure and Check Baseline

3.3.3.1	Configure processes and functions
3.3.3.2	Transport objects in QA environment
3.3.3.3	Test baseline
3.3.3.4	Document and correct inaccuracies
3.3.3.5	Detail business blueprint
3.3.3.6	Ensure completeness of detailed configuration

3.3.4 Prepare Baseline Acceptance

3.3.4.1	Prepare baseline scenarios
3.3.4.2	Collate agenda for baseline acceptance
3.3.4.3	Prepare meeting for baseline acceptance

3.3.5 Perform Baseline Acceptance

| 3.3.5.1 | Examine baseline scenarios |
| 3.3.5.2 | Check and accept baseline |

3.4 System Administration

3.4.1 Work Out System Test Plans

3.4.1.1	Work out test plan for system downtime scenarios
3.4.1.2	Work out throughput test plan
3.4.1.3	Work out stress test plan
3.4.1.4	Work out test plan for system administration
3.4.1.5	Work out print and fax test plan

3.4.2 Define Service Level Agreement

3.4.2.1	Set system downtime scenarios
3.4.2.2	Define disaster recovery procedures
3.4.2.3	Set up service level agreement

3.4.3 Set Up System Administration Functions

| 3.4.3.1 | Check procedure for client copy |
| 3.4.3.2 | Inspect daily check procedures |

3.4.3.3   Inspect transport procedures

3.4.3.4   Inspect backup and recovery procedures

3.4.4 Set Up Quality Assurance Environment

3.4.4.1   Install hardware for QA system

3.4.4.2   Inspect technical system environment

3.4.4.3   Install quality assurance system

3.4.4.4   Set up user master data for quality assurance

3.4.4.5   Save quality assurance system

3.4.4.6   Set up print services

3.4.4.7   Set up client administration and the transport system

3.4.5 Set Production System Design

3.4.5.1   Inspect estimates for workload and disk space

3.4.5.2   Design disk layout for the production system

3.4.6 Define System Management for the Production System

3.4.6.1   Set security concept for the production system

3.4.6.2   Define productive operation procedures

3.4.6.3   Define system administration for the production system

3.4.6.4   Define print environment for the production system

3.4.6.5   Define administration procedures for the production database

3.4.6.6   Crete operations manual for the R/3 System

3.4.7 Set Up Production Environment

3.4.7.1   Install hardware for the production system

3.4.7.2   Inspect the technical environment of the production system

3.4.7.3   Install production system

3.4.7.4   Install and configure network environment

3.4.7.5   Install desktop hardware and components

3.4.7.6   Save operating system and database

3.4.7.7	Install printers and configure print functions

3.5 Detail Configuration and Acceptance

3.5.1 Work Out Planning for Detail Configuration

3.5.1.1	Refine detail scope
3.5.1.2	Create configuration plan for detail scope
3.5.1.3	Determine test cases
3.5.1.4	Create test plan for detail scope
3.5.1.5	Assign resources
3.5.1.6	Schedule configuration workshops
3.5.1.7	Agree on planning for detail configuration

3.5.2 Hold Configuration Workshops (Cycle 1- n)

3.5.2.1	Hold workshop (cycle 1-n)
3.5.2.2	Document decisions on business processes (cycle 1-n)
3.5.2.3	Documents and clarify issues (cycle 1-n)

3.5.3 Configure and Check Detail Scope (Cycle n-1)

3.5.3.1	Configure processes and functions (cycle 1-n)
3.5.3.2	Transport objects in the QA environment (cycle 1-n)
3.5.3.3	Test detail configuration (cycle 1-n)
3.5.3.4	Ensure completeness of detail configuration

3.5.4 Prepare Detail Acceptance

3.5.4.1	Prepare detail acceptance scenarios
3.5.4.2	Collate agenda for detail acceptance
3.5.4.3	Prepare meeting on detail acceptance

3.5.5 Perform Detail Acceptance

3.5.5.1	Work through detail acceptance scenarios
3.5.5.2	Check and accept detail scope

3.6 Develop Data Transfer Programs

3.6.1 Work Out Data Transfer Procedures

3.6.1.1	Create detailed definition of data transfer
3.6.1.2	Create data transfer programs
3.6.1.3	Perform manual data transfer procedures

3.6.2 Test and Transport Date Transfer Programs

3.6.2.1	Define test procedure for data transfer
3.6.2.2	Test and check data transfer programs
3.6.2.3	Test results of data transfer
3.6.2.4	Transport Programs into the QA system

3.7 Develop Interface Programs for Applications

3.7.1 Create Interface Programs

3.7.1.1	Work out detailed definition of interfaces
3.7.1.2	Develop dialog interface programs
3.7.1.3	Develop background interfaces

3.7.2 Test and Transport Interface Programs

3.7.2.1	Define test procedures for interfaces
3.7.2.2	Test and check interface programs
3.7.2.3	Agree on the results of interface tests
3.7.2.4	Transport interface programs into the

3.8 Develop Enhancements

3.8.1 Work Out Enhancement Programs

3.8.1.1	Work out detailed definition of enhancement
3.8.1.2	Check approval
3.8.1.3	Create enhancements

3.8.2 Test and Transport Enhancement Programs

3.8.2.1	Define test procedures for enhancements
3.8.2.2	Test and check enhancement programs

3.14 Documentation and Training Materials for Users

3.14.1 Work Out Development Plan for User Documentation

3.14.1.1	Determine documentation requirements of users
3.14.1.2	Work out development plan for user documentation
3.14.1.3	Agree on development plan for user documentation

3.14.2 Work Out User Documentation

3.14.2.1	Hold workshop on user documentation and training
3.14.2.2	Create user documentation

3.14.3 Work Out Training Materials for Users

3.14.3.1	Collate training materials for users
3.14.3.2	Create training guidelines for trainers

3.14.4 Prepare User Training

3.14.4.1	Organize rooms, equipment, and logistics for training
3.14.4.2	Confirm list of participants

3.15 Implement Quality Check

3.15.1	Perform and accept quality check
3.15.1.1	Perform quality check
3.15.1.2	Accept implementation

Phase 4: Preparation for Production

4.1 Project Management Preparation for Production Phase

4.1.1 Project Status Meeting

4.1.1.1	Prepare status meetings
4.1.1.2	Hold status meetings
4.1.1.3	Check completion of assigned tasks
4.1.1.4	Correct project deviations
4.1.1.5	Adjust project plan

4.1.2 Steering Committee

4.1.2.1	Prepare steering committee meetings
4.1.2.2	Hold steering committee meetings
4.1.2.3	Check completion of assigned tasks

4.1.3 General Project Management

4.1.3.1	Change management
4.1.3.2	Hold team-building activities

4.2 User Training

4.2.1 Prepare User Training

4.2.1.1	Complete logistics for training
4.2.1.2	Set up environment for user training
4.2.1.3	Transport training data into the training environment

4.2.2 Hold User Training

4.2.2.1	Hold user training
4.2.2.2	Check user training

4.3 System Management

4.3.1 Set Up Administration of the Production System

4.3.1.1	Configure CCMS for the production environment
4.3.1.2	Configure print and spool administration for the production system
4.3.1.3	Train personnel for system administration

4.3.2 Perform System Tests

4.3.2.1	Perform throughput test
4.3.2.2	Perform stress test
4.3.2.3	Perform system administration tests
4.3.2.4	Perform disaster recovery test
4.3.2.5	Test backup and restore procedure
4.3.2.6	Perform print and fax tests

4.3.2.7   GoingLive (Check)

4.4 Detailed Planning for Cut-Over und Support

4.4.1 Work Out Planning for Cut-Over

4.4.1.1   Check procedure for data transfer

4.4.1.2   Create checklists for data transfer

4.4.1.3   Set readiness for production

4.4.1.4   Agree on cut-over planning

4.4.2 Work Out Detailed Planning for Production Support

4.4.2.1   Determine help desk procedures

4.4.2.2   Set up help desk

4.4.2.3   Name team for production support

4.4.2.4   name help desk personnel

4.4.2.5   Set a long-term support strategy

4.5 Cut-Over

4.5.1 Perform Cut-Over to the Production System

4.5.1.1   Transport into the production environment

4.5.1.2   Perform data transfer

4.5.1.3   Enter data manually

4.5.2 Final Acceptance for Productive Operations

4.5.2.1   Accept production system

4.5.2.2   Save production system

4.5.2.3   Ensure readiness of users

4.6 Quality Check: Preparation for Production

4.6.1 Perform and Accept Quality Check

4.6.1.1   Perform quality check

4.6.1.2   Accept preparation for production phase

Phase 5: Go-Live und Support

5.1 Production Support

5.1.1 Ready Production Support

5.1.1.1	Communicate problems and issues
5.1.1.2	Clarify issues and solve problems

5.1.2 Check Results of Productive Business Processes

5.1.2.1	Monitor daily and weekly transactions
5.1.2.2	Clarify issues
5.1.2.3	Confirm live environment

5.1.3 Optimize System Usage

5.1.3.1	Hold EarlyWatch meetings
5.1.3.2	Optimize technical environment
5.1.3.3	Optimize R/3 transactions

5.2 Activities After the Go-Live

5.2.1 Follow-Up Training

5.2.1.1	Hold training for advanced users
5.2.1.2	Train new employees

5.2.2 Define Long-Term Plans

5.2.2.1	Detail a technical upgrade plan
5.2.2.2	Work out detailed plans for additional process requirements
5.2.2.3	Convert plans to a long-term support strategy

5.2.3 Upgrade the Production Landscape

5.2.3.1	Perform a system upgrade
5.2.3.2	Perform a database upgrade
5.2.3.3	Perform release Customizing
5.2.3.4	Check user acceptance

5.2.4 Perform Continuing System Transactions

5.2.4.1   Perform daily transactions in the computer center

5.2.4.2   Perform daily transactions of database administration

5.2.4.3   Refine system administration

5.2.5 Project Review

5.2.5.1   Check and clarify open questions

5.2.5.2   Check business and project goals

5.2.5.3   Accept and close list of open questions

Glossary

Access sequence: Search strategy for the finding of a condition record. The access sequence determines the sequence in which the system gains access to condition records.

APO: Advanced Planned Optimizer: APO includes planning functions for the strategic, tactical and operational planning of logistics chains. The entire area of planning is covered by the integrated APO-modules. APO is provided with several integrated modules, which have access to a joint data basis.

Availability check: Inventory audit, which automatically ensues with every goods movement and which shall prevent that the posting inventory of physical inventory categories gets negative.

Batch: Partial amount of a material, which is processed separately from other partial amounts of the same material in the stock. Examples for batches are different production batches (e.g. according to color or medicaments), delivery lots or quality steps of a material.

Budget plan: Plan, which is subordinated to the project plan. It compares monthly planned costs to actual costs and marks differences.

Business area: A business area is a legally independent, organizational unit within a client, for which internal balance sheets can be generated. This takes place either on basis of the sales organization/sector or the delivering plant.

Business scenario: A sequence of business transactions, which have in common the dependency on an event or a period. Event controlled scenarios base upon a certain event as e.g. the placing of a customer order. Time related scenarios do not base upon certain events, but upon a period. The month-end closing, the standard costs evaluations, test runs and possible data reorganizations are such processes.

Business transaction: The business transaction presents the detailed instructions for the installation, the plan, the evaluations and the improvement of the R/3-solution and definition according to the business requirements.

Business Blueprint: It builds a structure for the understanding of the company business targets and the development of the written documentation for the processes necessary for the support of these targets.

Business Blueprint-document: The most important worked-out information of the Blueprint. The Business Blueprint Document is a written documentation of the results, in order to determine the requirements. The purpose of these documents is, to check, whether the requirements have been understood and communicated in the right way. The Business Blueprint also determines the exact project size.

Business Process Master List: An R/3- performance of the project size, which is in the realization phase developed to practicable business scenarios and R/3-transactions.

Business Process Procedure: Model for the starting definition for the development of working procedures and business-and test transactions.

Business Reengineering: Today's market requirements force companies to permanently improve the quality and activity size of their products. The most efficient method here for is, if taking into consideration the experiences of the practice, to completely consider anew and optimize structures and transaction. This procedure is called Business Reengineering. Business Reengineering and the connected concepts and methods have become the paragon of programs, which require on grounds of their radical, structural change, a fundamental renewal of present company- and organization structures. Because of the reflection and radical reformation of the central business processes, the Business Reengineering can lead to significant improvements.

Business-to-Business (B2B): For the procedure of the scenario B2B, during which a company orders via net, receives the invoice and also pays, SAP offers the New Dimension Product "Business to Business Procurement". This product enables the companies to perform business processes on basis of Internet, in the sense of the Supply Chain Management. The B2B scenario allows the control and procedure of business processes starting with the order requirement over the state inquiry to the payment of an invoice.

Business-to-Business Procedure: The application B2B Procurement is one of the first applications being delivered in the scope of the SAP-Initiative EnjoySAP. It makes the procurement of C-materials, which are not planned by the ERP-system, possible, as e.g. office supply, operating supply and services. The SAP

B2B Procurement consequently enables an open, completely company-overlapping procurement and controls the entire procurement processes from the order requirement to the invoice payment. With the help of electronic catalogs, the users can procure most easily goods and services directly from their work place. The purchase department is therefore released and can concentrate on the strategic purchase e.g. for the component supplier management or price negotiations.

Business-to-Consumer (B2C): This scenario is also called Electronic Retailing. In so-called "Shopping Malls" various types of consumer goods can be offered to the customer. The scenario can be displayed with the help of the R/3-component SAP Online Store.

Calculation basis: The calculation basis determines the origin of the data, taken as basis in the pricing for a condition type.

Example: Delivery, delivery position, dispatch element, freight cost position.

Calculation scheme: Regulation for the pricing. It determines the conditions, being permitted for the pricing of a business transaction, and establishes the sequence in which they are to be considered.

Change management: It refers to the usage of the R/3-project conversion from one environment into another. This movement can be processed out of the business-/company perspective (change of the company or the business type) or out of the IS-perspective (the company changes the system or changes to a current system release). In connection with the project management the Change Management refers to changes of size, budget, life period and resources.

Classification: Division of a position in the sales document, referring date and amount. If the total amount of a position can only be covered by four partial amounts, the system automatically generates four divisions, to which the particular amount and date is determined.

Collective invoice: Settlement form, with which one invoice per period is created for a customer.

Company code: The smallest organizational unit of the external accounting, for which a complete, compact accounting can be displayed. This includes all accounting relevant events and the creation of all verifications for a legal individual company closing.

Computer Added Test Tool (CATT): Universally usable SAP-test tool, with which operational transactions can be summarized in repeatable test procedures and be automated.

Condition: Agreement about prices, surcharges and advance payment, taxes and so on, depending on selected influence factors (as e.g. deliverer, customer, customer group, material, activity) determined and valid within a certain period.

Condition record: Data record, in which conditions and possibly also additional conditions are displayed. Following conditions can be registered in the condition record:

- Prices

- Surcharges and advance payments

- Taxes

- Information

Condition type: Recognition, which defines features and characteristics of a condition. In the pricing e.g. rebate of net price or discount of gross price are differentiated from each other by different condition types.

Consignment: Consignment means that a deliverer provides material to a customer, which is stored by the customer. The deliverer remains proprietor of the material until the customer withdraws a good off the consignment stock. Reliability towards the deliverer is the result. The invoice is placed referring agreed periods, e.g. monthly. The contract partners can agree that the customer transfers the material of the remaining consignment stock into his own stock, after a certain time.

Consignment fill-up: It is the dispatch of goods in the consignment stock of the customer. Is there no special stock for the customer, it is automatically generated with the posting of the goods issue. The corresponding amount is posted in the customer special stock. The consignment stock stays property of the deliverer.

Consignment pick-up: It is the redelivery of materials off the customer consignment stock. The good has up to now not been billed to the customer.

Consignment return: The consignment return is the return of goods having been withdrawn off the customer consignment stock and therefore have already been invoiced. Since the good has returned in the property of the original company, the transaction is invoice relevant. The customer receives a credit memo for the returned good.

Consignment stock: Stock, which is made available by the deliverer but stays in his property until it is withdrawn and used or being transferred into the own, evaluated stock.

Consignment withdrawal: Consignment withdrawal is the property transfer of good to the customer and the invoicing. At the closing of the month a withdrawal message is made in case of withdrawal of the consignment stock by the customer. The corresponding amount is subtracted with posting both from the customer special stock and the evaluated total stock. Now the good is invoiced.

Contact: A data record filed in the system, which contains information about interactions with customers. Possible contact types are e.g. customer visits, telephone calls, conferences or lectures.

Contact partner: Person of the customer company, with whom the sales-or marketing department of the delivering company maintains business contacts.

Contract: Long-term agreement with deliverers, which is fulfilled through single call-ups.

Credit limit: The credit limit defines the credit quantity limit in the currency of a credit control area.

Credit memo requirement: Reference document for the creation of a credit memo. A credit memo requirement is created in the purchase, on demand of the customer.

Creditworthiness check: With every order generating a creditworthiness of the customer should be checked. Here it is important, that the check is not statically, but dynamically processed. Referring a dynamic check, the dates for invoice receipts can be forecasted. Therefore, a possible exceeding the overdraft, caused by the entered and checked order, can be recognized. A limit for the sum of open claims and the value of up to the moment not invoiced orders is generally filed in the customer master record.

Customer delivery: Summarization of sales material, which can be jointly delivered. A customer delivery is the basis for the internal organization of the dispatch activities, resulting in the delivery of the good to the goods recipient.

Customer group: A group of customers freely chosen, for the pricing or statistical purposes. A customer group can comprise e.g. wholesale trade or retail industry.

Customer hierarchy: Display of complex structures. The regional grouping of for example purchase associations can be performed as customer hierarchy. For a customer hierarchy price information are filed, which are valid for all customers of the hierarchy.

Customer inquiry: Customer request to a sales organization to place a customer quotation.

Customer order: Contract agreement between a sales organization and an ordering party about the delivery of materials or the rendering of services on determined conditions.

Customer quotation: Quotation of a sales organization towards a customer for the delivery of materials or the rendering of services referring determined conditions.

Customizing: Applications being processed in the implementation of the system. The customizing enables following:

- To be able to install SAP-functions quick, safely and inexpensively.

- To be able to adjust standard functions, in order to meet the business requirements of the company.

- To be able to document and administrate implementation- and adjustment phases in a simple tool for the project control.

Customizing-object: Summarization of the customizing-tables and views, which form, referring to operational criteria, a unit and therefore have to be maintained and transported together.

Deadline check: In regard of the order creation and-check it is necessary to check, whether the customer desired delivery date stated in the order could be met. The meeting of deadlines can extreme important for the customer ties. It can be realized through the usage of a deadline check tool. This tool processes the so-called ATP (Available-to-Promise)-check. Hereby it is inspected, whether a delivery promise can be given and met.

Debit memo requirement: Reference document for the creation of a debit memo on grounds of price- or quantity deviations or other complaints.

Delivering plant: Plant, from which goods are to be sent to a customer.

Delivery: Sales document for the procedure of good delivery. The delivery serves as basis for all dispatch activities:

- Disposition of the demand amounts

- Picking

- Generating of dispatch documents

- Creation of dispatch elements

- Transport

- Invoicing

Delivery stock: Labor stock for the creation of deliveries. The labor stock comprises all orders and dispatch schedules to be delivered in a certain period.

Dispatch center: Organization unit of logistics handling the dispatch process. The dispatch center divides the responsibilities in the company referring the dispatch type, the necessary dispatch auxiliaries and transport means.

Dispatch document: Document serving the display of business transactions in the dispatch. There are following dispatch documents:

- Delivery

- Material document including the goods issue data

- Delivery groups

Dispatch element: Summarization of materials, which are packed together with a dispatch auxiliary for a specific date.

Dispatch notification: Acceptance of the delivery date for a certain amount of ordered goods by a deliverer.

Disposition: Different techniques, using the bill of material, stock data and the production plan, in order to calculate the material demands.

Document: Proof of a business transaction in the sales procedure. Hereby, three different documents are distinguished:

- Sales document

- Dispatch document

- Invoice document

Document flow: Display of the successive documents of a business transaction. A document flow can consist e.g. of a quotation, an order, a delivery and an invoice.

E-commerce: In the scope of the E-commerce, two main-scenarios matter the most:

- Business-to-Business (B2B) and

- Business-to-Consumer (B2C)

External data: Not SAP-data. Data, coming from an external company.

Freight costs (document): An object, in which the freight cost relevant data are comprised. The freight cost document enables the reference to the components of a transport, for which are separate freight calculation shall be made. Within the freight cost document the freight costs are calculated, posted and transferred to FI and CO.

Goods issue: Term of the inventory management, which characterizes a reduction of stock on grounds of a goods withdrawal or-issue or a delivery to a customer. In order to ensure short material-and information lead times in the goods issue area, as well as low total costs, a variety of features is to be considered. The necessary transshipment-and transport facilities are influenced most of all by condition, shape and weight of the materials due to transport. The temporal allocation of shipments determines the required personnel-and transport capacity.

Goods recipient: Person or company, to whom/which the good is delivered. The good recipient is not necessarily identical with the ordering party, invoice recipient or payer.

Implementation: Change of an existing system and the kind of business procedures, in order to process with the help of R/3-system a part or the entire business.

Implementation assistant: PC-based application, including various tools. The IA produces a tight connection between documents, to which it is made access through the Accelerated SAP Roadmap.

Implementation Guide (IMG): The implementation guide is available for the processing of the system setups. With the help of the implementation guide, system setups can be proceeded for all applications of the R/3-system. The implementation guide is divided in application areas. Within the application areas, the working steps are structured referring the sequence of their processing.

Implementation strategy: Procedure for the implementation of the R/3-system. It can be a Big Bang-procedure (all applications at one time), level estimate (a limited quantity of applications or business processes at one time) or a procedure on plant-or sector level.

Incompleteness protocol: Protocol, in which all data missing in a sales document are listed. For every sales type it can be separately defined, which data are necessary for the completeness of the document.

Inquiry: The request of a customer to a sales organization to placing a quotation.

Integration: The result of different functionalities in the system, which are influenced by single transactions and also affect these. Examples:

The generating of a customer order can result in the inspection of the inventory through the material administration and effect a material planning.

Deliverer data, which are used both by the customer and the purchase.

Issue: Unforeseen activity, project or business situation, influencing the business-and project targets and delaying the project time schedule. It can result in changes of the project size, the budget, the time schedule and the resources.

Invoice: Sales document, with which goods delivery or services are passed to the company account.

Invoice deadline: Date, at which a delivery is due to invoice. In some companies invoices are periodically handled, so that all deliveries can be summarized in one collective invoice and billed together. As soon as the next invoice deadline determined is reached, the summarized orders and deliveries are displayed in the invoice stock and ready to be invoiced.

Invoice list: Invoicing type, with which all invoices of a period are summarized for a payer. Additional advance payments as e.g. Delcredere can be granted on the total value of an invoice list. Invoices of an invoice list can be single or collective invoices.

Invoice recipient: Person or company receiving an invoice for deliveries made or services rendered. An invoice recipient is not automatically identical with the payer, who has to pay the invoice at last.

Invoice split: Creation of several invoices of the reference document (order or delivery). A possible application for the invoice split is the invoice separation referring goods groups.

Invoicing: Generic term for bills, credit-and debit memos and reversals. An invoice consists of a document header with data, applied for the entire document, and a particular quantity of document positions.

Knowledge Corner: Detailed description of R/3-processes and functions, accompanied by adjustment auxiliaries for certain application areas. These are applied during the creation of requirements and during the definition.

Loading point: Place within a dispatch position, at which the good is loaded.

Master data: The basic data of a company and their quality characterize the fundamental basis for the successful procedure of business transactions. Within the operational standard software package SAP R/3 these basic data are filed as so called "Master data". The master data in the R/3-system are stored in central position and form the foundation of all business processes in Sales. It refers to data relating to individual objects, which remain over a long period unchanged. Master data include information, which are used for similar objects in the same way.

The master data of a deliverer contain name, address and bank information.

The master data of a user in the SAP-system include user name, permissions, standard printer.

Material: Product, substance or object with which it is traded or the production starts, used or created; a material can also be a service.

Material exclusion: Sales restrictions, which exclude specific materials from the sale to a customer. The system blocks in the sales documents to this customer the enter of materials, which are excluded from the sale to this customer.

Material finding: Automatic finding of a material master record during the sales document creation with any key, instead of the own material number. The key given in the sales document, with which a material can be found, can for example be a customer individual material number or the EAN-number of the material.

Material listing: Sales restriction, which controls the sale of certain material to the customer. The customer can only buy the material included in the material listing. The system blocks the enter of materials not included in the material listing in a sales document of the customer.

One-time customer: Name for a collective customer master record, which is applied for the procedure of transactions with different customers, who are no regular customers of the deliverer. If a transaction for a one-time customer is registered, the customer data must be manually added.

Order: Request of a purchase organization to a deliverer or a plant, to deliver a certain quantity of products to a certain time or to render services.

Order combination: Combination of entire orders, individual order positions of different orders or partial deliveries of individual order positions in a delivery. Order combination in a delivery is only possible, when it has been approved for a customer in the customer master data in general or manually for single orders in an order document header.

Ordering party: Business partner, who causes deliveries or services. The ordering party can also fulfill the function of a payer, invoice recipient and goods recipient.

Order procedure: For the control of the entire goods flow in the goods distribution and the coordination of all individual transactions, the usage of an order procedure system is absolutely necessary. Only this enables the interaction of persons, equipments and procedures within certain structures, by providing suitable information. Consequently, the order procedure of a sales system is set over the operative procedures of the product flow. Hereby, the development of different partial systems, which process specific tasks as master data record, the goods control of the warehouses, goods accounting and so on, serves for the acceleration and rationalization of the information flow performance. The rating of such a system is very high, since only through the availability of premature, solid information flexible and fast sales can be realized.

Outline agreement: The generic term for contracts and delivery notes. The outline agreement is a long-term agreement with a deliverer about the delivery of materials or the rendition of services to determined conditions.

Partial payment: Payment, with which the invoiced amount is only partially settled.

Partial delivery: Goods receipt quantity, being smaller than the order amount minus underdelivery tolerance.

Partner: Unit within or outside the own organization, which is of commercial interest and with which in the scope of the business transaction, connections can be built up. A partner can be a natural or legal person.

Payer: Business partner, who makes payments for delivered goods or rendered services. The payer is not automatically identical with the goods receiver.

Picking: Withdrawal and summarization of certain partial amounts (materials) off the stock on grounds of demand information of the sales or production. The picking can be processed by means of transport orders or picking lists.

Plant: Organization unit of logistics, which subdivides the company referring the production, procurement and disposition. The materials are produced respectively provided in a plant.

Position: Part of a document including information about the goods to be delivered or the activities.

Predefined client: The predefined client is a sequence of definitions, which consists of the frequently applied US-Customizing (e.g. US-orientated account tables, measurement units and forms). The basic processes MM, SD and FI/CO are established and run from the first day of the clients activation.

Presales procedure: This scenario describes the definition of the master data, which are the prerequisite for the procedure of sales scenarios for industrial customers.

Price elements: Price elements for example are:

- Prices
- Advance payment

- Surcharges

- Freights

- Taxes

Procedure model: The procedure model shows the structure of the R/3-implementation objects.

Process: An R/3-transaction, which effects the creation or change of a business object or a condition.

Procurement logistics:

A high and flexible reaction capacity towards the customer requirements depends to a great extent on the provision of material components of external component suppliers. Here, the procurement tasks have to be fulfilled. The procurement logistics has to ensure, that in industrial plants the, of the production required, raw- materials and operating supplies are provided in sufficient amounts and in the right time and that in trading companies the ordered products can be provided with the minimization of the warehouse and transport costs.

Production logistics: It takes care for the optimal provision of the semi products in the single steps of the production process. The emphasis lies here in the physical performance of the material flow.

Production preparation: Phase 4 of the Accelerated SAP Roadmap. In this phase the closing tests, trainings of users and the preparations of the cutovers (both for the data and the system) are terminated for a productive environment.

Productive system: A system, which contains a company-specific business process and which is applied for the standard operating. In the productive system, productive data are entered.

Profile Generator: The Profile Generator supports the authorization administrator with the creation of an authorization concept at the company site. The Profile Generator automatically creates an authorization profile on grounds of an application component sequence chosen by the administrator. Then the administrator can adjust and change the authorizations included in the profile with the help of special maintenance transactions of the Profile Generator.

Project Estimator (PE): Presale tool, which generates a baseline estimation of the time, the resources and costs, which are combined with a specific R/3-implementation. Information referring the size, the company level, the SAP-team expertise and the level of complexity of the business processes, are only a few questions the Project Estimator deals with. Proceeding the Project Estimator, a UPB-size document and a project plan can be generated.

Project plan: Plan including three components: Budget plan, resource plan and a routing.

Project preparation: Phase 1 of the Accelerated SAP Roadmap, in which the project team of the company is arranged and trained in the Accelerated SAP-method, a project plan on highest level is effected, the project standard and procedures are determined and a hardware order is checked and executed.

Purchase organization: Organization unit of logistics, which divides the company in regard of the requirements of the purchase. A purchase organization procures material or services, negotiates conditions and is responsible for these business transactions.

Q&A-data base: The Q&A- database is a repository of all questions and the corresponding answers of the company. These questions and answers are required for the definition of the business requirements and for the development of business solutions in regard of the R/3-reference module and the R/3-system. Business processes, technical questions, organizational questions and questions referring the definition (and naturally also the answers), which form the resource for the creation of the Business Blueprint, are included.

Realization: Phase 3 of the Accelerated SAP Roadmap, in which the Blueprint requirements of the R/3-system are defined and tested. The business requirements of the company are evaluated and transformed into the future R/3-business solution. This transformation process is an iterative evaluation-and improvement process, in which every business process runs through 3 phases:

- Acceptance. The business requirements of the company are evaluated in order to determine, how they should be dealt with in the R/3-system.

- Revision. Every business process is required, until the optimal Business Project Procedure is gained.

- Termination. Every BBP is optimized together with the practicable definition, according to the business processes and the information requirements.

Rebate: A price rebate, which is subsequently guaranteed to the customer by the deliverer. The amount of the rebate depends in general on the turnover, which is gained with the customer throughout a fixed period.

Rebate agreement: Agreement between the deliverer and the customer about the granting of rebates. A rebate agreement includes discount relevant information (e.g. rebate basis-and size, rebate recipients and validity period).

Reference document: Document, out of which data are copied into another.

Return: A goods return delivery by the customer. A return is planned by means of a return plan. The entry of a good is registered with the help of a return delivery and the good posted to the stock.

Returnable packaging: Packaging material or transport auxiliaries for the transport or the storage of materials. The returnable packaging is delivered to the customer with the goods transport and is to be returned.

Revenue account finding: Determination of revenue accounts, to which prices, surcharges and advance payments are posted. The revenue accounts are automatically determined with the help of conditions.

Roadmap: A sequence of processes, which determine the steps necessary for the R/3-implementation. The 5 phases of the Accelerated SAP Roadmap (depicted by means of a street graphic) are:

1.   Project preparation

2.   Business Blueprint

3.   Realization

4.   Production preparation

5.   Go-Live and Support

Route: Distance to be covered between a starting and a closing point. A route can consist of several sections.

R/3-system: The R/3-system consists of:

- A central authority, which offers the services DUEBMGS (Dialogue, Update, Enqueue, Background procedure, Messages, Gateway and Spool)

- A data base instance
- Optional dialogue authority
- Optional PC-Frontends

Sale from stock to the consignment taker: The sales from stock to consignment taker includes all necessary processes of the sales to sell products from stock to the customers, who at first lay the products in their consignment storage. In particular, these are customer orders with or without reference to quotations or outline agreements, the delivery and dispatch procedures as well as the invoicing. Further more the scenario contains processes for the handling of customer complaints, subsequent payments as rebates or commissions, as well as of returnable packaging and loose goods.

Sales area: The sales area is a combination of sales organization, sales path and sector. The sales area is consequently a unit, which comprises three aspects:

- Sales orientated aspect (sales organization)
- Customer orientated aspect (sales path)
- Product orientated aspect (sector)

Sales document: Document for the display of business processes in the sales. It has to be differentiated between following sales documents:

- Inquiry
- Quotation
- Order
- Outline agreement
- Returns
- Credit memo requirements
- Debit memo requirements

Sales document (SD): A document in the system, which displays a transaction in sales/distribution. It is differentiated between following sales documents:

- Sales documents
- Dispatch documents
- Invoice documents

Sales document type: Recognition for the different control of various sales documents. Document types effect, that different business transactions as e.g. standard orders and credit memo requirements with subsequent credit memos are variably processed in the system.

Sales information system (SD): The sales information system included in the SAP R/3-system serves, as part of the logistics information system (LIS), for the rapid identification of existing problem areas by means of solid ratios and for the analysis of the cause of their development. Basis of this ratios are the operational transactions, which are regularly processed and are cumulated for the purpose of an evaluation.

Sales logistics: The sales logistics is the link between the production and the sales part of the company. Task of the sales logistics is to perform, control and check the goods flows of a company referring the sales market. Target is to deliver the demanded good in time to the customer by minimizing the storage- and transport costs.

Sales organization: Organization unit of the logistics, which divides the company according to the sales requirements. A sales organization is responsible for the sales/distribution of material and services. With this component it is controlled, how the company divides the markets on grounds of geographical and sector specific criteria. Every business transaction is processed by exactly one sales organization.

Sales path: The sales path marks the way, on which the goods reach the customer.

SAP-Data Archive: The SAP-data archive is the standard functionality of the R/3 system. It serves for the consistent transfer of application data, which are not any longer required in the online-operating. This leads in general to an improved performance of the database and therefore to quicker answer times in the dialogue operating. Because of the reduced database volume, the archiving expense and the requirement of additional hardware components are simultaneously decreased. The SAP-data archiving enables a quicker and secured data transfer off the database and the storage of archive data, on which you can have further access.

SAP Employee Self Service (ESS): The SAP Employee Self Service-applications bind the employee closer and release the personnel division in regard of routine tasks. ESS is a solid solution, with which the employees can take over responsibility for their own data, without attending a special R/3-training. With SAP ESS the occasional user can directly maintain Human Resources management data in the R/-system, by means of a browser, which can be applied intuitively.

SAP Online Store: The SAP Online Store is an Internet application component for the electronic retailing between companies and consumers as well as between companies. Manufacturers, mail-order houses, wholesale trade and retail industry can directly offer their products worldwide by means of the World Wide Web. The particularity of the SAP Online Store is the smoothing and homogeneous integration of data to be processed in the R/3-system.

Scenario: A sequence of business transactions, having in common a dependency of an event or a period. Event-controlled scenarios base upon a specific event as e.g. the placing of a customer order. Time depending scenarios base not upon a certain event, but upon the period. The month-end closing, standard costs evaluation, test runs ands possible data reorganizations belong to these processes.

Scheduling: The scheduling comprises the calculation start-and closing dates of orders and transaction in the orders.

Scheduling agreement: Form of an outline agreement. The scheduling agreement is a long-term agreement with a deliverer or customer, which is assigned for the creation and regular actualization of time planning for the delivery of partial amounts of the material of every position. The time planning is developed by the creation of schedule lines.

Sector: An organizational unit built with regard of the sales competence or the profit responsibility of saleable materials or services

Service render: Business partner, rendering certain services to the company. This can be e.g. a forwarding agent, an insurance, a freight supplier and similar.

Settlement date: Date, to which the position is registered in the financial accounting or cost accounting. Further more the settlement date enters into the tax determination.

Stock: Term of the material management for a part of the current assets. It comprises the amount of raw material, operating supplies, semi products and finished products as well as trading goods.

Stock transfer: The release of materials from a certain stock and their storage into another storage location. Transfers can take place either within a plant as well as between two plants. The release from stock and the transfer to stock can be posted in the system by one or two steps.

Storage location: Organization unit, which enables the differentiation of material stocks within a plant.

Supply Chain Cockpit (SCC): The Supply Chain Cockpit is a graphical instrument table for the modeling, performance/depiction, planning and control of the Supply Chain. This tool enables the user to, on the one hand, have a view on the logistics chain as such and shows on the other hand the control possibilities of the planning-and scheduling processes, underlying the logistics chain.

Supply Chain Management: In the scope of this new management concept, the idea of the lean management is transferred to the entire value added chain. Hereby the elimination of company internal extravagances along this value added chain is strived for.

Tank car procedure: The sale from stock of tank cars to industrial customers includes all necessary processes of the Sales. The particularity of loose goods, being sold in tank cars, is considered through the direct connection between the customer order and the filling of the tank car before the delivery.

Third-party deal: Trade with goods not concerning the stock since being directly delivered to the customer by the deliverer.

Third-party order processing: With a third- party order the customer orders products from a company, which does not produce by its own, respectively does not keep the product in stock. Here, the delivery of the materials ordered by the customer is not processed by the company receiving the order. The order is handed over instead to an external deliverer, who sends the material directly to the customer. The invoice however, the customer receives from that company, which accepted the order; the deliverer on his part invoices that company accepting the order.

Transaction code: Sequence of four symbols characterizing a SAP-transaction. To call up a transaction in R/3,you have to enter the transaction code in the instruction field. SM31 is e.g. the code for the transaction table maintenance.

Transport: The transport of materials and goods serves the overcoming of special distances. Referring the external transport in the scope of the Sales Logistics, the transport ensues from the company to the customer.

Transportation: Organizational and planning activity in the approaches of a transport.

Transportation position: An organizational unit of Logistics, which is responsible for the planning and procedure of transport activities. The transportation position divides the responsibilities in the company e.g. according to the type of the transport, carrier or the regional departments.

Transport management: Route planning and transport cost management can be processed by means of the transport management. This component of the Logistics Executive System comprises functionalities for following activities:

- Dispatch scheduling

- Route planning

- Freight cost calculation

- Transport procedure and -inspection

Transport procedure: Target of the transport procedure is the processing of all steps necessary, to send a transport to the customer. Transport tasks e.g. is the weighing of a transport, the loading and the goods issue posting.

Transport section: Sections include information about the geographical conditions of a transport. These sections for example are:

- **Distance**: Connection between a starting- and closing point.

- **Transshipment position**: Place, at which the transport is transferred from one transport medium to another.

- **Limiting point**: Point, at which the transport crosses a border.

Transport type: A transport type is a specific procedure variant for a transport. The transport type includes all necessary control features for a transport.

Example: Single transport, collective transport, main route.

Warehouse management: The warehouse management system is a component of the Logistics Executive System. But it can also be used as independent component. A warehouse management system supports among others following functions:

- Administration of storage structures and –establishments

- Survey of the stock movement

- Transfer to/release from stock of dangerous goods

References

/Alpar96/ Alpar, P.: Kommerzielle Nutzung des Internet, Berlin 1996.

/Balzert98/ Balzert, H.: Lehrbuch der Software - Technik: Software - Management, Software - Qualitätssicherung, Unternehmensmodellierung, Heidelberg, Berlin 1998.

/Balzert98/ Helmut Balzert: Lehrbuch der Software-Technik Band 2, 1998, Spektrum Akademischer Verlag

/Bartels80/ Bartels, H.G.: Logistik: In: Albers, W. (Hrsg.): Handwörterbuch der Wirtschaftswissenschaft, Stuttgart et al. 1980.

/Busher87/ Busher, J.R., Tyndall, G.R.: Logistics Excellence, in Management Accounting August 1987

/Beims95/ Beims, H. D.: Praktisches Software Engineering, Vorgehen, Methoden, Werkzeuge, München, Wien 1995

/Berthold99/ Berthold A., Mende U., Schuster H.: SAP Business Workflow – Konzept, Anwendung, Entwicklung, 1.Auflage, München 1999

/Bieberstein93/ Bieberstein, N.: CASE-Tools: Auswahl, Bewertung, Einsatz, München, Wien 1993.

/Bloech84/ Bloech, J.: Problembereiche der Logistik. In: Schriften zur Unternehmensführung Heft 32, Wiesbaden 1984

/Boehm81/ Boehm, B. W.: Software Engineering, Englewood Cliffs 1981.

/Brand 98/ Hartwig Brand: SAP R/3-Einführung mit ASAP. Technische Implementierung von SAP R/3 planen und realisieren, 1. Unveränderter Nachdruck, München 1999

/Buck-Emden99/ Buck-Emden, R..: Die Technologie des SAP-Systems: Basis für betriebswirtschaftliche Anwendungen , 4. durchges. u. korr. Auflage, München 1999.

/Curran98/ Curran T., Keller, G., : SAP R/3 Business Blueprint – Business Reegineering mit R/3-Referenzprozessen, 1.Auflage, München 1999.

/Frick95/ Frick, A.: Der Softwareentwicklungsprozeß. Ganzheitliche Sicht, München 1995.

/**Geiss, Soltysiak98**/ Geiß, M., Soltysiak, R.: SAP R/3 dynamisch einführen, 1. Auflage, Bonn 1998

/**Hammer96**/ Hammer, M., Champy, J.: Business Reengineering: Die Radikalkur für das Unternehmen, Frankfurt 1996

/**Heinemann98**/ Heinemann, G.: Dynamisierung im Absatzkanal: ECR – ein Allheilmittel?, in: Absatzwirtschaft, Ausgabe 10/1998

/**Hohberger96**/ Hohberger, P: Erfahrungen in der Unternehmenspraxis, in: Office Management, Ausgabe 7-8/1996

/**Jacob99**/ Jacob O., Uhink H.-J.: SAP R/3 im Mittelstand, 1. Auflage, Braunschweig/Wiesbaden 1998

/**Jäger93**/ Jäger, E. et al.: Die Auswahl zwischen alternativen Implementierungen von Geschäftsprozessen im einem Standardsoftwarepaket am Beispiel eines KFZ-Zulieferes, In: Wirtschaftsinformatik 35, 1993.

/**Jamin94**/ Jamin, K.-W.: Das Software-Lexikon, 3., akt. Auflage, Renningen-Malmsheim 1994.

/**Keller99**/ Keller, G. & Partner.: SAP R/3 prozeßorientiert anwenden, 3. Erweiterte Auflage, München 1999.

/**Knollmayer99**/ Knollmayer G., Mertens P., Zeier A..: Supply Chain-Management auf der Basis von SAP-Systemen – Perspektiven der Auftragsabwicklung für Industriebetriebe, 1.Auflage, Berlin-Heidelberg 2000

/**Körsgen98**/ Körsgen F.: SAP R/3 Vertrieb: Fallstudien Anwendung und Customizing, 1. Auflage, Berlin 1999.

/**Mertens93**/ Mertens P.: Integrierte Informationverarbeitung, Wiesbaden 1993.

/**Nieschlag94**/ Nieschlag, R. Dichtl, E., Hörschgen, H. 17., neu bearb. Aufl., Berlin 1994

/**Oberniedermaier99**/ Oberniedermaier G., Geiß, M., : SAP R/3-Systeme effizient testen, 1. Auflage, München 1999

/**Paulk95**/ Mark C. Paulk, Charles V. Weber, Bill Curtis, Mary Beth Chtissis: The Capability Maturity Model: Guidelines for Improving the Software Process. Reading u. a. 1995

/**Perez98**/ Perez M., Hildenbrand A., Matzke B., Zencke P..: Geschäftsprozesse im Internet im SAP R/3, 1. korr. Nachdruck, Bonn, Reading, Mass. 1998.

/Pfohl85/ Pfohl, H.C.; Logistiksysteme, Berlin, Heidelberg, New York, Tokyo 1985.

/Pomberger96/ Pomberger, G., Blaschek, G.: Software Engineering, 2., überarb. Auflage, München, Wien 1996.

/Poth73/ Poth, L.G.: Praxis der Marketing-Logistik, Heidelberg 1973.

/Managermagazin98/ Rieker, J.: Die drei von der Baustelle, in Managermagazin 4/98.

/Börse Online97/ Rüppel, W.: SAP – einfach die erfolgreichste Aktie, in Börse Online Nr. 46/97.

/Roell85/ Roell, J.S.: Das Informations- und Entscheidungssystem der Logistik, Frankfurt, Bern, New York 1985.

/SCHEer90/ Scheer, A.-W.: Wirtschaftsinformatik – Informationssysteme im Industriebetrieb, Berlin 1990

/Schneider97/ Schneider, H.-J.: Lexikon Informatik und Datenverarbeitung, 4., akt. u. durchges. Auflage, München, Wien 1997.

/Schwarting86/ Schwarting, C.:Optimierung der ablauforganisatorischen Gestaltung von Kommissioniersystemen, München 1986.

/Schulte95/ Schulte, C.: Logistik – Wege zur Optimierung des Material- und Informationsflusses, 2. überarbeitete und erweiterte Auflage, München 1995.

/Stahlknecht95/ Stahlknecht, P.; Einführung in die Wirtschaftsinformatik, 7., vollst. überarb. und erw. Auflage, Berlin 1995.

/Sommerville92/ Ian Sommerville: Software Engineering. 4. Auflage, Wokingham, Großbritannien u. a. 1992

/Stapleton97/ Jennifer Stapleton: Dynamic Systems Development Method. The Method In Practice. Harlow, Großbritannien 1997

/Teufel99/ Teufel T, Röhricht J., Willems P.: SAP R/3 Prozeßanalyse mit Knowledge Maps, 1. Auflage, München 1999.

/Thome96/ Thome, R., Hufgard, A.: Continuous System Engineering, 1. Auflage, Würzburg 1996.

/Thome97/ Thome, R., Schinzer, H.: Marktüberblick, in Thome R., Schinzer H. (Hrsg.) Electronic Commerce: Anwendungsbereiche und Potentiale der digitalen Geschäftsabwicklung, München 1997.

/**Trauboth93**/ Trauboth, H.: Software-Qualitätssicherung, München, Wien 1993.

/**Traumann76**/ Traumann, P.: Markting-Logistik in der Praxis, Main 1976.

/**Ungermann96**/ Ungermann, C., Tesch, F.-J., Stolpe, B., Weissert, M.: Qualitätsmanagement bei der Softwareerstellung: Leitfaden für die Umsetzung der DIN EN ISO 9000, Düsseldorf 1996

/**VDI92**/ VDI-Gemeinschaftsausschuß CIM: Qualitätssicherung, Düsseldorf 1992

/**Wenzel96**/ Wenzel, P.: Betriebswirtschaftliche Anwendungen des integrierten Systems SAP R/3, 2., verb. u. erw. Auflage, Braunschweig, Wiesbaden 1996.

/**Witt94**/ Witt, J.: Absatzmanagement Gabler, Wiesbaden 1994.

/**Xu95**/ Xu, Z.-Y.: Prinzipien des Entwurfs und der Realisierung eines Organisationsinformationssystems, Heidelberg 1995.

/**Zöller91**/ Zöller, H.: Wiederverwendbare Software-Bausteine in der Automatisierung, Düsseldorf 1991.

Index

T

Tank car procedure 113
  Description 113
Target of logistics
  Logistical performance 17
  Optimization of the logistical profits 17
  Profitability factors of logistics 17
  Logistical costs 17
Targets of SCM
  Operative targets 278
Targets of the use of standard applications
  30
Test 318
Test administrator 318
Test catalogue 318
Test evaluation 318
Test execution 318
Test group 208
Test package 318
Test plans 318
Test process 318
Test rule 208
Test scenarios 318
Test Workbench 294, 318
Testers 318

Testing Transactions 296
Third party order processing 64
  Definition 64
  Procedure 67
Threshold value analysis 198
Timebox-approach 248
Transport management 290
Trend analysis 198

U

Usage 115
Usage fields of CATT with the
  implementation of the SAP R/3-module
  SD 295
User training 299

V

Value added tax 178

W

Warehouse management 79, 289